FREE Test Taking Tips DVD Offer

To help us better serve you, we have developed a Test Taking Tips DVD that we would like to give you for FREE. **This DVD covers world-class test taking tips that you can use to be even more successful when you are taking your test.**

All that we ask is that you email us your feedback about your study guide. Please let us know what you thought about it – whether that is good, bad or indifferent.

To get your **FREE Test Taking Tips DVD**, email freedvd@studyguideteam.com with "FREE DVD" in the subject line and the following information in the body of the email:

    a. The title of your study guide.

    b. Your product rating on a scale of 1-5, with 5 being the highest rating.

    c. Your feedback about the study guide. What did you think of it?

    d. Your full name and shipping address to send your free DVD.

If you have any questions or concerns, please don't hesitate to contact us at freedvd@studyguideteam.com.

Thanks again!

Common Core Math Grade 4 Textbook & Workbook

Common Core 4th Grade Math Workbook & Practice Test Questions

Test Prep Books

Table of Contents

Quick Overview

As you draw closer to taking your exam, effective preparation becomes more and more important. Thankfully, you have this study guide to help you get ready. Use this guide to help keep your studying on track and refer to it often.

This study guide contains several key sections that will help you be successful on your exam. The guide contains tips for what you should do the night before and the day of the test. Also included are test-taking tips. Knowing the right information is not always enough. Many well-prepared test takers struggle with exams. These tips will help equip you to accurately read, assess, and answer test questions.

A large part of the guide is devoted to showing you what content to expect on the exam and to helping you better understand that content. Near the end of this guide is a practice test so that you can see how well you have grasped the content. Then, answers explanations are provided so that you can understand why you missed certain questions.

Don't try to cram the night before you take your exam. This is not a wise strategy for a few reasons. First, your retention of the information will be low. Your time would be better used by reviewing information you already know rather than trying to learn a lot of new information. Second, you will likely become stressed as you try to gain large amount of knowledge in a short amount of time. Third, you will be depriving yourself of sleep. So be sure to go to bed at a reasonable time the night before. Being well-rested helps you focus and remain calm.

Be sure to eat a substantial breakfast the morning of the exam. If you are taking the exam in the afternoon, be sure to have a good lunch as well. Being hungry is distracting and can make it difficult to focus. You have hopefully spent lots of time preparing for the exam. Don't let an empty stomach get in the way of success!

When travelling to the testing center, leave earlier than needed. That way, you have a buffer in case you experience any delays. This will help you remain calm and will keep you from missing your appointment time at the testing center.

Be sure to pace yourself during the exam. Don't try to rush through the exam. There is no need to risk performing poorly on the exam just so you can leave the testing center early. Allow yourself to use all of the allotted time if needed.

Remain positive while taking the exam even if you feel like you are performing poorly. Thinking about the content you should have mastered will not help you perform better on the exam.

Once the exam is complete, take some time to relax. Even if you feel that you need to take the exam again, you will be well served by some down time before you begin studying again. It's often easier to convince yourself to study if you know that it will come with a reward!

Test-Taking Strategies

1. Predicting the Answer

When you feel confident in your preparation for a multiple-choice test, try predicting the answer before reading the answer choices. This is especially useful on questions that test objective factual knowledge or that ask you to fill in a blank. By predicting the answer before reading the available choices, you eliminate the possibility that you will be distracted or led astray by an incorrect answer choice. You will feel more confident in your selection if you read the question, predict the answer, and then find your prediction among the answer choices. After using this strategy, be sure to still read all of the answer choices carefully and completely. If you feel unprepared, you should not attempt to predict the answers. This would be a waste of time and an opportunity for your mind to wander in the wrong direction.

2. Reading the Whole Question

Too often, test takers scan a multiple-choice question, recognize a few familiar words, and immediately jump to the answer choices. Test authors are aware of this common impatience, and they will sometimes prey upon it. For instance, a test author might subtly turn the question into a negative, or he or she might redirect the focus of the question right at the end. The only way to avoid falling into these traps is to read the entirety of the question carefully before reading the answer choices.

3. Looking for Wrong Answers

Long and complicated multiple-choice questions can be intimidating. One way to simplify a difficult multiple-choice question is to eliminate all of the answer choices that are clearly wrong. In most sets of answers, there will be at least one selection that can be dismissed right away. If the test is administered on paper, the test taker could draw a line through it to indicate that it may be ignored; otherwise, the test taker will have to perform this operation mentally or on scratch paper. In either case, once the obviously incorrect answers have been eliminated, the remaining choices may be considered. Sometimes identifying the clearly wrong answers will give the test taker some information about the correct answer. For instance, if one of the remaining answer choices is a direct opposite of one of the eliminated answer choices, it may well be the correct answer. The opposite of obviously wrong is obviously right! Of course, this is not always the case. Some answers are obviously incorrect simply because they are irrelevant to the question being asked. Still, identifying and eliminating some incorrect answer choices is a good way to simplify a multiple-choice question.

4. Don't Overanalyze

Anxious test takers often overanalyze questions. When you are nervous, your brain will often run wild causing you to make associations and discover clues that don't actually exist. If you feel that this may be a problem for you, do whatever you can to slow down during the test. Try taking a deep breath or counting to ten. As you read and consider the question, restrict yourself to the particular words used by the author. Avoid thought tangents about what the author *really* meant, or what he or she was *trying* to say. The only things that matter on a multiple-choice test are the words that are actually in the question. You must avoid reading too much into a multiple-choice question, or supposing that the writer meant something other than what he or she wrote.

5. No Need for Panic

It is wise to learn as many strategies as possible before taking a multiple-choice test, but it is likely that you will come across a few questions for which you simply don't know the answer. In this situation, avoid panicking. Because most multiple-choice tests include dozens of questions, the relative value of a single wrong answer is small. Moreover, your failure on one question has no effect on your success elsewhere on the test. As much as possible, you should compartmentalize each question on a multiple-choice test. In other words, you should not allow your feelings about one question to affect your success on the others. When you find a question that you either don't understand or don't know how to answer, just take a deep breath and do your best. Read the entire question slowly and carefully. Try rephrasing the question a couple of different ways. Then, read all of the answer choices carefully. After eliminating obviously wrong answers, make a selection and move on to the next question.

6. Confusing Answer Choices

When working on a difficult multiple-choice question, there may be a tendency to focus on the answer choices that are the easiest to understand. Many people, whether consciously or not, gravitate to the answer choices that require the least concentration, knowledge, and memory. This is a mistake. When you come across an answer choice that is confusing, you need to give it extra attention. A question might be confusing because you do not know the subject matter to which it refers. If this is the case, don't eliminate the answer before you have affirmatively settled on another. When you come across an answer choice of this type, set it aside as you look at the remaining choices. If you can confidently assert that one of the other choices is correct, you can leave the confusing answer aside. Otherwise, you will need to take a moment to try to better understand the confusing answer choice. Rephrasing is one way to tease out the sense of a confusing answer choice.

7. Your First Instinct

Many people struggle with multiple-choice tests because they overthink the questions. If you have studied sufficiently for the test, you should be prepared to trust your first instinct once you have carefully and completely read the question and all of the answer choices. There is a great deal of research suggesting that the mind can come to the correct conclusion very quickly once it has obtained all of the relevant information. At times, it may seem to you as if your intuition is working faster even than your reasoning mind. This may in fact be true. The knowledge you obtain while studying may be retrieved from your subconscious before you have a chance to work out the associations that support it. Verify your instinct by working out the reasons that it should be trusted.

8. Key Words

Many test takers struggle with multiple-choice questions because they have poor reading comprehension skills. Quickly reading and understanding a multiple-choice question requires a mixture of skill and experience. To help with this, try jotting down a few key words and phrases on a piece of scrap paper. Doing this concentrates the process of reading and forces the mind to weigh the relative importance of the question's parts. In selecting words and phrases to write down, the test taker thinks about the question more deeply and carefully. This is especially true for multiple-choice questions that are preceded by a long prompt.

9. Subtle Negatives

One of the oldest tricks in the multiple-choice test writer's book is to subtly reverse the meaning of a question with a word like *not* or *except*. If you are not paying attention to each word in the question, you can easily be led astray by this trick. For instance, a common question format is, "Which of the following is...?" Obviously, if the question instead is, "Which of the following is not....?," then the answer will be quite different. Even worse, the test makers are aware of the potential for this mistake and will include one answer choice that would be correct if the question were not negated or reversed. A test taker who misses the reversal will find what he or she believes to be a correct answer and will be so confident that he or she will fail to reread the question and discover the original error. The only way to avoid this is to practice a wide variety of multiple-choice questions and to pay close attention to each and every word.

10. Reading Every Answer Choice

It may seem obvious, but you should always read every one of the answer choices! Too many test takers fall into the habit of scanning the question and assuming that they understand the question because they recognize a few key words. From there, they pick the first answer choice that answers the question they believe they have read. Test takers who read all of the answer choices might discover that one of the latter answer choices is actually *more* correct. Moreover, reading all of the answer choices can remind you of facts related to the question that can help you arrive at the correct answer. Sometimes, a misstatement or incorrect detail in one of the latter answer choices will trigger your memory of the subject and will enable you to find the right answer. Failing to read all of the answer choices is like not reading all of the items on a restaurant menu: you might miss out on the perfect choice.

11. Spot the Hedges

One of the keys to success on multiple-choice tests is paying close attention to every word. This is never more true than with words like *almost, most, some*, and *sometimes*. These words are called "hedges", because they indicate that a statement is not totally true or not true in every place and time. An absolute statement will contain no hedges, but in many subjects, like literature and history, the answers are not always straightforward or absolute. There are always exceptions to the rules in these subjects. For this reason, you should favor those multiple-choice questions that contain hedging language. The presence of qualifying words indicates that the author is taking special care with his or her words, which is certainly important when composing the right answer. After all, there are many ways to be wrong, but there is only one way to be right! For this reason, it is wise to avoid answers that are absolute when taking a multiple-choice test. An absolute answer is one that says things are either all one way or all another. They often include words like *every, always, best*, and *never*. If you are taking a multiple-choice test in a subject that doesn't lend itself to absolute answers, be on your guard if you see any of these words.

12. Long Answers

In many subject areas, the answers are not simple. As already mentioned, the right answer often requires hedges. Another common feature of the answers to a complex or subjective question are qualifying clauses, which are groups of words that subtly modify the meaning of the sentence. If the question or answer choice describes a rule to which there are exceptions or the subject matter is complicated, ambiguous, or confusing, the correct answer will require many words in order to be expressed clearly and accurately. In essence, you should not be deterred by answer choices that seem excessively long. Oftentimes, the author of the text will not be able to write the correct answer without offering some qualifications and modifications. Your job is to read the answer choices thoroughly and completely and to select the one that most accurately and precisely answers the question.

13. Restating to Understand

Sometimes, a question on a multiple-choice test is difficult not because of what it asks but because of how it is written. If this is the case, restate the question or answer choice in different words. This process serves a couple of important purposes. First, it forces you to concentrate on the core of the question. In order to rephrase the question accurately, you have to understand it well. Rephrasing the question will concentrate your mind on the key words and ideas. Second, it will present the information to your mind in a fresh way. This process may trigger your memory and render some useful scrap of information picked up while studying.

14. True Statements

Sometimes an answer choice will be true in itself, but it does not answer the question. This is one of the main reasons why it is essential to read the question carefully and completely before proceeding to the answer choices. Too often, test takers skip ahead to the answer choices and look for true statements. Having found one of these, they are content to select it without reference to the question above. Obviously, this provides an easy way for test makers to play tricks. The savvy test taker will always read the entire question before turning to the answer choices. Then, having settled on a correct answer choice, he or she will refer to the original question and ensure that the selected answer is relevant. The mistake of choosing a correct-but-irrelevant answer choice is especially common on questions related to specific pieces of objective knowledge, like historical or scientific facts. A prepared test taker will have a wealth of factual knowledge at his or her disposal, and should not be careless in its application.

15. No Patterns

One of the more dangerous ideas that circulates about multiple-choice tests is that the correct answers tend to fall into patterns. These erroneous ideas range from a belief that B and C are the most common right answers, to the idea that an unprepared test-taker should answer "A-B-A-C-A-D-A-B-A." It cannot be emphasized enough that pattern-seeking of this type is exactly the WRONG way to approach a multiple-choice test. To begin with, it is highly unlikely that the test maker will plot the correct answers according to some predetermined pattern. The questions are scrambled and delivered in a random order. Furthermore, even if the test maker was following a pattern in the assignation of correct answers, there is no reason why the test taker would know which pattern he or she was using. Any attempt to discern a pattern in the answer choices is a waste of time and a distraction from the real work of taking the test. A test taker would be much better served by extra preparation before the test than by reliance on a pattern in the answers.

FREE DVD OFFER

Don't forget that doing well on your exam includes both understanding the test content and understanding how to use what you know to do well on the test. We offer a completely FREE Test Taking Tips DVD that covers world class test taking tips that you can use to be even more successful when you are taking your test.

All that we ask is that you email us your feedback about your study guide. To get your **FREE Test Taking Tips DVD**, email freedvd@studyguideteam.com with "FREE DVD" in the subject line and the following information in the body of the email:

- The title of your study guide.
- Your product rating on a scale of 1-5, with 5 being the highest rating.
- Your feedback about the study guide. What did you think of it?
- Your full name and shipping address to send your free DVD.

Introduction to the Common Core Math Grade 4 Test

Function of the Test

The purpose of the Common Core Math Grade 4 test to assess fourth grade students' mastery of the mathematics standards that have been identified as those appropriate for the grade level. The material tested on the Common Core Math Grade 4 should be addressed by in-class mathematics education and the curriculum designed by the school the student attends. Achieving a high score on the test demonstrates that a given student understands the mathematics concepts deemed important for grade-level academic success and that instructional delivery of the skills and concepts was effective.

Test Administration

Over 40 states have adopted the Common Core Standards. Each one of these states develops and uses their own version of a Common Core Math test and as such, administration varies from state to state. The test is usually split into two or three testing sessions that take place during school hours in place of normal instructional. Students who finish early are usually instructed to read quietly or do other independent work.

Test Format

The test format for Common Core Math tests varies from state to state; however, the material tested is consistent across states. The number and type of questions varies. The 2018 New York State Grade 4 Math test contained 45 questions: 38 multiple-choice questions, 6 short-response questions, and 1 extended response. Testing was divided into two sessions. In contrast, Louisiana's test was administered in three sessions. It contained 60 multiple-choice questions (36 without the use of a calculator and 24 where a calculator was permitted) and 3 constructed-response question.

The Grade 4 exam focuses on three primary areas: fluently multiplying and dividing multi-digit numbers, operations with fractions and identification of equivalent fractions, and the analysis of and classification of geometric figures based on their attributes.

The following table shows the domains of mathematics assessed on the exam the number of standards tested for each:

Domain	Number of Standards Tested
Operations and Algebraic Thinking	3
Number and Operations in Base Ten	2
Number and Operations: Fractions	3
Measurement and Data	3
Geometry	1

Scoring

Because each state has the latitude to design and administer their own version of the test, scores cannot be easily compared between states.

Operations and Algebraic Thinking

Using the Four Operations with Whole Numbers to Solve Problems

Interpreting a Multiplication Equation as a Comparison

Multiplication can be completed using place value. When a number is multiplied times 10, the number shifts over one place value to the left, and a 0 is entered in the ones place. For example, $1{,}235 \times 10 = 12{,}350$. Similarly, when a number is multiplied times 100, the entire number is shifted over two place values to the left, and a 0 is entered in both the ones and tens places. For example, $15{,}634 \times 100 = 1{,}563{,}400$. This same technique can be used to multiply single digit numbers times factors of 10 and 100. For instance, 5×300 can be thought of as $5 \times 3 \times 100 = 15 \times 100 = 1{,}500$.

Multiplication involves taking multiple copies of one number. The sign designating a multiplication operation is the x symbol. The result is called the **product**. For example, $9 \times 6 = 54$. Multiplication means adding together one number in the equation as many times as the number on the other side of the equation:

$$9 \times 6 = 9 + 9 + 9 + 9 + 9 + 9 = 54$$

$$9 \times 6 = 6 + 6 + 6 + 6 + 6 + 6 + 6 + 6 + 6 = 54$$

The product, 54, of 9×6 can be thought of as six groups of nine, or nine groups of six.

Consider the problem 3×4. A model can be used to visualize these groups.

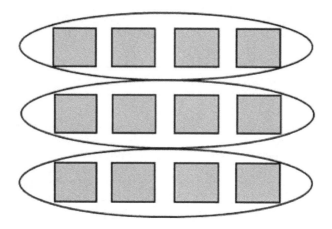

In this first graphic (above), there are three groups of four squares. The product of 3×4 is 12 because there are 12 total boxes. The same problem is displayed in the alternative grouping below:

Here there are still 12 boxes that model 3×4, but there are four groups with three boxes per group.

Multiplication also adheres to certain properties of operations. When reviewing calculations consisting of more than one operation, the order in which the operations are performed affects the resulting answer. Consider $5 \times 2 + 7$. Performing multiplication then addition results in an answer of 17 ($5 \times 2 = 10$; $10 + 7 = 17$). However, if the problem is written $5 \times (2 + 7)$, the order of operations dictates that the operation inside the parenthesis must be performed first. The resulting answer is 45 ($2 + 7 = 9$, then $5 \times 9 = 45$).

The order in which operations should be performed is remembered using the acronym PEMDAS. PEMDAS stands for parenthesis, exponents, multiplication/division, and addition/subtraction. Multiplication and division are performed in the same step, working from left to right with whichever comes first. Addition and subtraction are performed in the same step, working from left to right with whichever comes first.

Consider the following example: $8 \div 4 + 8(7 - 7)$. Performing the operation inside the parenthesis produces $8 \div 4 + 8(0)$ or $8 \div 4 + 8 \times 0$. There are no exponents, so multiplication and division are performed next from left to right resulting in: $2 + 8 \times 0$, then $2 + 0$. Finally, addition and subtraction are performed to obtain an answer of 2. Now consider the following example: $6x3 + 3^2 - 6$. Parentheses are not applicable. Exponents are evaluated first, $6 \times 3 + 9 - 6$. Then multiplication/division forms $18 + 9 - 6$. At last, addition/subtraction leads to the final answer of 21.

With any number times one (for example, $8 \times 1 = 8$) the original amount does not change. Therefore, one is the **multiplicative identity**. For any whole number a, $1 \times a = a$. Also, any number multiplied times zero results in zero. Therefore, for any whole number a, $0 \times a = 0$.

Multiplication also follows the commutative property. The order in which multiplication is calculated does not matter. For example, 3 x 10 and 10 x 3 both equal 30. Ten sets of three apples and three sets of ten apples both equal 30 apples. The **commutative property of multiplication** states that for any whole numbers a and b, $a \times b = b \times a$. Multiplication also follows the associative property because the product of three or more whole numbers is the same, no matter what order the multiplication is completed. The **associative property of multiplication** states that for any whole numbers *a, b,* and *c,* $(a \times b) \times c = a \times (b \times c)$. For example, $(2 \times 3) \times 4 = 2 \times (3 \times 4)$.

The **distributive property of multiplication over addition** is an extremely important concept that appears in algebra. It states that for any whole numbers *a, b,* and *c,* it is true that $a \times (b + c) = (a \times b) + (a \times c)$. Because multiplication is commutative, it is also true that $(b + c) \times a = (b \times a) + (c \times a)$. For example, $100 \times (3 + 2)$ is the same as $(100 \times 3) + (100 \times 2)$. Both result in 500.

Multiplying or Dividing to Solve Word Problems Involving Multiplicative Comparison

In solving multi-step problems, the first step is to line up the available information. Then, try to decide what information the problem is asking to be found. Once this is determined, construct a strip diagram to display the known information along with any information to be calculated. Finally, the missing information can be represented by a **variable** (a letter from the alphabet that represents a number) in a mathematical equation that the student can solve.

For example, Delilah collects stickers and her friends gave her some stickers to add to her current collection. Joe gave her 45 stickers, and Aimee gave her 2 times the number of stickers that Joe gave Delilah. How many stickers did Delilah have to start with, if after her friends gave her more stickers, she had a total of 187 stickers?

In order to solve this, the given information must first be sorted out. Joe gives Delilah 45 stickers, Aimee gives Delilah 2 times the number Joe gives (2×45), and the end total of stickers is 187.

A strip diagram represents these numbers as follows:

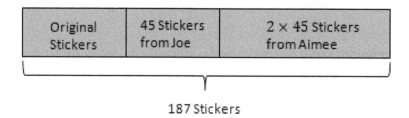

187 Stickers

The entire situation can be modeled by this equation, using the variable *s* to stand for the original number of stickers:

$$s + 45 + (2 \times 45) = 187.$$

Solving for *s* would give the solution, as follows:

$$s + 45 + 90 = 187$$

$$s + 135 = 187$$

$$s + 135 - 135 = 187 - 135$$

$$s = 52 \text{ stickers.}$$

Word problems take concepts you learned in the classroom and turn them into real-life situations. Some parts of the problem are known and at least one part is unknown. There are three types of instances in which something can be unknown: the starting point, the change, or the final result. These can all be missing from the information they give you.

For solving problems with unknown factors, it is often easiest to set up an array to visualize the grouping of the information. In these problems, set up the initial numbers in uniformly sized groups, so the solution can be determined by inspection of the grouping.

Find the missing number (?) in the following equation:

$$? \times 5 = 35$$

Knowing that one of the factors is to be multiplied is 5 allows the groupings to be made in sets of five columns. In this case, there 5 columns of items are created, until the desired number (35) is reached.

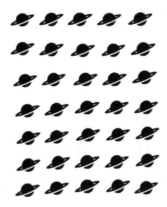

Here, the number of 35 is reached with the seventh row of items. Therefore, the missing factor is 7.

$$5 \times 7 = 35$$

The same problem could be demonstrated with the equation:

$$5 \times ? = 35$$

This would simply require the information to be grouped into five rows, and items added evenly until the desired number (35) is reached.

Again, the solution is:

$$5 \times 7 = 35.$$

This demonstrates the commutative property of multiplication by showing the missing factor could be the number of rows or the number of columns, and yet result in the same solution.

Solving Multi-step Word Problems

Calculations relating to real-world expenses rarely have whole number solutions. It is good practice to be able to make one-step calculations that model real-world situations. For example, Amber spends $54 on pet food in a month. If there are no increases in price for the next 11 months, how much will Amber spend on pet food during those 11 months?

Set up the multiplication problem to see how much $54 of food times 11 is:

```
   54
 × 11
   54
+540
 594
```

Amber would spend $594 on pet food over the next 11 months.

As another example, what if Amber had $650; how many months would this last her for pet food expenses if they were $54 per month?

Set the problem up as a division problem to see how many times 54 could divide into 650. The answer will give the number of complete months the expenses could be covered.

$$
\begin{array}{r}
12 \\
54\overline{)650} \\
-\ 54 \\
\hline
110 \\
-\ 108 \\
\hline
2
\end{array}
$$

The remainder does not represent a full month of expenses. Therefore, Amber's $650 would last for a full 12 months of pet food expenses.

In addition to these examples, there are a variety of real-world situations in which multiplication and division are used to solve problems. The tables below display some of the most common scenarios.

	Unknown Product	Unknown Group Size	Unknown Number of Groups
Equal groups	There are 5 students, and each student has 4 pieces of candy. How many pieces of candy are there in all? $5 \times 4 =?$	14 pieces of candy are shared equally by 7 students. How many pieces of candy does each student have? $7 \times ? = 14$ Solved by inverse operations $14 \div 7 =?$	If 18 pieces of candy are to be given out 3 to each student, how many students will get candy? $? \times 3 = 18$ Solved by inverse operations $18 \div 3 =?$

	Unknown Product	Unknown Factor	Unknown Factor
Arrays	There are 5 rows of students with 3 students in each row. How many students are there? $5 \times 3 =?$	If 16 students are arranged into 4 equal rows, how many students will be in each row? $4 \times ? = 16$ Solved by inverse operations $16 \div 4 =?$	If 24 students are arranged into an array with 6 columns, how many rows are there? $? \times 6 = 24$ Solved by inverse operations $24 \div 6 =?$

	Larger Unknown	Smaller Unknown	Multiplier Unknown
Comparing	A small popcorn costs $1.50. A large popcorn costs 3 times as much as a small popcorn. How much does a large popcorn cost? $1.50 \times 3 =?$	A large soda costs $6 and that is 2 times as much as a small soda costs. How much does a small soda cost? $2 \times ? = 6$ Solved by inverse operations $6 \div 2 =?$	A large pretzel costs $3 and a small pretzel costs $2. How many times as much does the large pretzel cost as the small pretzel? $? \times 2 = 3$ Solved by inverse operations $3 \div 2 =?$

Concrete objects can also be used to solve one- and two-step problems involving multiplication and division. Tools such as tiles, blocks, beads, and hundred charts are used to model problems. For example, a hundred chart (10×10) and beads can be used to model multiplication. If multiplying 5 by 4, beads are placed across 5 rows and down 4 columns producing a product of 20. Similarly, tiles can be used to model division by splitting the total into equal groups. If dividing 12 by 4, 12 tiles are placed one at a time into 4 groups. The result is 4 groups of 3. This is also an effective method for visualizing the concept of remainders.

Remainders in Division Problems
If a given total cannot be divided evenly into a given number of groups, the amount left over is the **remainder**. Consider the following scenario: 32 textbooks must be packed into boxes for storage. Each

box holds 6 textbooks. How many boxes are needed? To determine the answer, 32 is divided by 6, resulting in 5 with a remainder of 2. A remainder may be interpreted three ways:

- Add 1 to the quotient
  How many boxes will be needed? Six boxes will be needed because five will not be enough.

- Use only the quotient
  How many boxes will be full? Five boxes will be full.

- Use only the remainder
  If you only have 5 boxes, how many books will not fit? Two books will not fit.

The Reasonableness of Results
When solving math word problems, the solution obtained should make sense within the given scenario. The step of checking the solution will reduce the possibility of a calculation error or a solution that may be *mathematically* correct but not applicable in the real world. Consider the following scenarios:

A problem states that Lisa got 24 out of 32 questions correct on a test and asks to find the percentage of correct answers. To solve the problem, a student divided 32 by 24 to get 1.33, and then multiplied by 100 to get 133 percent. By examining the solution within the context of the problem, the student should recognize that getting all 32 questions correct will produce a perfect score of 100 percent. Therefore, a score of 133 percent with 8 incorrect answers does not make sense and the calculations should be checked.

A problem states that the maximum weight on a bridge cannot exceed 22,000 pounds. The problem asks to find the maximum number of cars that can be on the bridge at one time if each car weighs 4,000 pounds. To solve this problem, a student divided 22,000 by 4,000 to get an answer of 5.5. By examining the solution within the context of the problem, the student should recognize that although the calculations are mathematically correct, the solution does not make sense. Half of a car on a bridge is not possible, so the student should determine that a maximum of 5 cars can be on the bridge at the same time.

Once a result is determined to be logical within the context of a given problem, the result should be evaluated by its nearness to the expected answer. This is performed by approximating given values to perform mental math. Numbers should be rounded to the nearest value possible to check the initial results.

Consider the following example: A problem states that a customer is buying a new sound system for their home. The customer purchases a stereo for $435, 2 speakers for $67 each, and the necessary cables for $12. The customer chooses an option that allows him to spread the costs over equal payments for 4 months. How much will the monthly payments be?

After making calculations for the problem, a student determines that the monthly payment will be $145.25. To check the accuracy of the results, the student rounds each cost to the nearest ten (440 + 70 + 70 + 10) and determines that the total is approximately $590. Dividing by 4 months gives an approximate monthly payment of $147.50. Therefore, the student can conclude that the solution of $145.25 is very close to what should be expected.

When rounding, the place-value that is used in rounding can make a difference. Suppose the student had rounded to the nearest hundred for the estimation. The result $(400 + 100 + 100 + 0 =$

600; $600 \div 4 = 150$) will show that the answer is reasonable, but not as close to the actual value as rounding to the nearest ten.

Factors and Multiples

Factors and Multiples

The **factors** of a number are all integers that can be multiplied by another integer to produce the given number. For example, 2 is multiplied by 3 to produce 6. Therefore, 2 and 3 are both factors of 6. Similarly, $1 \times 6 = 6$ and $2 \times 3 = 6$, so 1, 2, 3, and 6 are all factors of 6. Another way to explain a factor is to say that a given number divides evenly by each of its factors to produce an integer. For example, 6 does not divide evenly by 5. Therefore, 5 is not a 6.

Multiples of a given number are found by taking that number and multiplying it by any other whole number. For example, 3 is a factor of 6, 9, and 12. Therefore, 6, 9, and 12 are multiples of 3. The multiples of any number are an infinite list. For example, the multiples of 5 are 5, 10, 15, 20, and so on. This list continues without end. A list of multiples is used in finding the least common multiple, or LCM, for fractions when a common denominator is needed. The denominators are written down and their multiples listed until a common number is found in both lists. This common number is the LCM.

Prime and Composite Numbers

Whole numbers are classified as either prime or composite. A prime number can only be divided evenly by itself and one. For example, the number 11 can only be divided evenly by 11 and one; therefore, 11 is a prime number. A helpful way to visualize a prime number is to use concrete objects and try to divide them into equal piles. If dividing 11 coins, the only way to divide them into equal piles is to create 1 pile of 11 coins or to create 11 piles of 1 coin each. Other examples of prime numbers include 2, 3, 5, 7, 13, 17, and 19.

A composite number is any whole number that is not a prime number. A composite number is a number that can be divided evenly by one or more numbers other than itself and one. For example, the number 6 can be divided evenly by 2 and 3. Therefore, 6 is a composite number. If dividing 6 coins into equal piles, the possibilities are 1 pile of 6 coins, 2 piles of 3 coins, 3 piles of 2 coins, or 6 piles of 1 coin. Other examples of composite numbers include 4, 8, 9, 10, 12, 14, 15, 16, 18, and 20.

To determine if a number is a prime or composite number, the number is divided by every whole number greater than one and less than its own value. If it divides evenly by any of these numbers, then the number is composite. If it does not divide evenly by any of these numbers, then the number is prime. For example, when attempting to divide the number 5 by 2, 3, and 4, none of these numbers divide evenly. Therefore, 5 must be a prime number.

Prime Factorization

Prime factorization breaks down each factor of a whole number until only prime numbers remain. All composite numbers can be factored into prime numbers. For example, the prime factors of 12 are 2, 2, and 3 ($2 \times 2 \times 3 = 12$). To produce the prime factors of a number, the number is factored and any composite numbers are continuously factored until the result is the product of prime factors only. A factor tree, such as the one below, is helpful when exploring this concept.

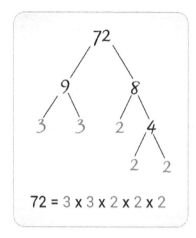

$$72 = 3 \times 3 \times 2 \times 2 \times 2$$

Generating and Analyzing Patterns

Generating a Number or Shape Pattern that Follows a Given Rule

Patterns are an important part of mathematics. Identifying and understanding how a group or pattern is represented in a problem is essential for being able to expand this process to more complex problems. A simple input-output table can model a pattern that pertains to a specific situation or equation. These can then be utilized in other areas in math, such as graphing.

For example, for every 1 parakeet the pet store sells, it sells 5 goldfish. Using the following equation to model this situation, fill in numbers missing in the input-output table, to show the total number of pets sold by the store.

Equation:

Total number of pets sold (t) = number of parakeets (p) + number of parakeets (p) × 5 goldfish

$$t = p + (p \times 5)$$
$$t = 6p$$

p	t
1	6
2	12
3	
4	24
5	

The missing numbers are 18 and 30.

This can also be shown by using an equation. If 3 is put in for p, it would look as follows:

$$t = 6 \times 3$$

$$t = 18$$

If 5 is put in for p, it would look as follows:

$$t = 6 \times 5$$

$$t = 30$$

The completed table would appear as follows:

p	t
1	6
2	12
3	18
4	24
5	30

Identifying Apparent Features of the Pattern that Were Not Explicit in the Rule Itself

Recognizing features of patterns is helpful because it becomes easier to determine the rule. Additionally, if the rule is already known, noting other features of the pattern that are not necessarily explicit in the rule itself can be used to ensure that the pattern is extended correctly or that the right subsequent number or shape is produced because it should also follow that feature.

For instance, given the sequence of numbers $7, 14, 21, 28, 35, \ldots$, the next number in the sequence would be 42. This is because the sequence lists all multiples of 7, starting at 7. This is the rule for the pattern. However, it can also be determined that the numbers generated in the pattern alternate between odd and even because an odd number (7) is being added to each previous value. Moreover, each odd-placed value in the pattern (the first value is 7, the third value is 21, the fifth value is 35, etc.) is odd, and every even-placed value is even (the second value is 14, the fourth value is 28, the sixth value is 42, etc.). Therefore, if a question asks for the tenth value in the pattern, you will know that number should be even because 10 is an even number. If you end up getting an odd number, you will know you have made a mistake and should go back and check your work.

Numbers and Operations in Base 10

Generalizing Place Value Understanding for Multi-Digit Whole Numbers

Interpreting the Value of Each Place-Value Position

In accordance with the base-10 system, the value of a digit increases by a factor of ten each place it moves to the left. For example, consider the number 7. Moving the digit one place to the left (70), increases its value by a factor of 10 ($7 \times 10 = 70$). Moving the digit two places to the left (700) increases its value by a factor of 10 twice ($7 \times 10 \times 10 = 700$). Moving the digit three places to the left (7,000) increases its value by a factor of 10 three times ($7 \times 10 \times 10 \times 10 = 7,000$), and so on.

Conversely, the value of a digit decreases by a factor of ten each place it moves to the right. (Note that multiplying by $\frac{1}{10}$ is equivalent to dividing by 10). For example, consider the number 40. Moving the digit one place to the right (4) decreases its value by a factor of 10 ($40 \div 10 = 4$). Moving the digit two places to the right (0.4), decreases its value by a factor of 10 twice ($40 \div 10 \div 10 = 0.4$) or ($40 \times \frac{1}{10} \times \frac{1}{10} = 0.4$). Moving the digit three places to the right (0.04) decreases its value by a factor of 10 three times ($40 \div 10 \div 10 \div 10 = 0.04$) or ($40 \times \frac{1}{10} \times \frac{1}{10} \times \frac{1}{10} = 0.04$), and so on.

Representing Values Using Expanded Notation and Numerals

The number system that is used consists of only ten different digits or characters. However, this system is used to represent an infinite number of values. The place value system makes this infinite number of values possible. The position in which a digit is written corresponds to a given value. Starting from the decimal point (which is implied, if not physically present), each subsequent place value to the left represents a value greater than the one before it. Conversely, starting from the decimal point, each subsequent place value to the right represents a value less than the one before it.

The names for the place values to the left of the decimal point are as follows:

...	Billions	Hundred-Millions	Ten-Millions	Millions	Hundred-Thousands	Ten-Thousands	Thousands	Hundreds	Tens	Ones

*Note that this table can be extended infinitely further to the left.

The names for the place values to the right of the decimal point are as follows:

Decimal Point (.)	Tenths	Hundredths	Thousandths	Ten-Thousandths	...

*Note that this table can be extended infinitely further to the right.

When given a multi-digit number, the value of each digit depends on its place value. Consider the number 682,174.953. Referring to the chart above, it can be determined that the digit 8 is in the ten-thousands place. It is in the fifth place to the left of the decimal point. Its value is 8 ten-thousands or 80,000. The digit 5 is two places to the right of the decimal point. Therefore, the digit 5 is in the hundredths place. Its value is 5 hundredths or $\frac{5}{100}$ (equivalent to .05).

A number is written in expanded form by expressing it as the sum of the value of each of its digits. The expanded form of 2,356,471.89, which is written with the highest value first down to the lowest value, is expressed as: 2,000,000 + 300,000 + 50,000 + 6,000 + 400 + 70 + 1 + .8 + .09 7,000 + 600 + 30 + 1 + .4 + .02.

When verbally expressing a number, the integer part of the number (the numbers to the left of the decimal point) resembles the expanded form without the addition between values. In the above example, the numbers read "seven thousand six hundred thirty-one." When verbally expressing the decimal portion of a number, the number is read as a whole number, followed by the place value of the furthest digit (non-zero) to the right. In the above example, 0.42 is read "forty-two hundredths." Reading the number 7,631.42 in its entirety is expressed as "seven thousand six hundred thirty-one and forty-two hundredths." The word *and* is used between the integer and decimal parts of the number.

Comparing and Ordering Whole Numbers
Rational numbers are any number that can be written as a fraction or ratio. Within the set of rational numbers, several subsets exist that are referenced throughout the mathematics topics. Counting numbers are the first numbers learned as a child. Counting numbers consist of 1,2,3,4, and so on. Whole numbers include all counting numbers and zero (0,1,2,3,4,…). Integers include counting numbers, their opposites, and zero (…,-3,-2,-1,0,1,2,3,…). Rational numbers are inclusive of integers, fractions, and decimals that terminate, or end (1.7, 0.04213) or repeat ($0.136\overline{5}$).

Placing numbers in an order in which they are listed from smallest to largest is known as **ordering**. Ordering numbers properly can help in the comparison of different quantities of items.

When comparing two numbers to determine if they are equal or if one is greater than the other, it is best to look at the digit furthest to the left of the decimal place (or the first value of the decomposed numbers). If this first digit of each number being compared is equal in place value, then move one digit to the right to conduct a similar comparison. Continue this process until it can be determined that both numbers are equal or a difference is found, showing that one number is greater than the other. If a number is greater than the other number it is being compared to, a symbol such as > (greater than) or < (less than) can be utilized to show this comparison. It is important to remember that the "open mouth" of the symbol should be nearest the larger number.

For example:

1,023,100 compared to 1,023,000

First, compare the digit farthest to the left. Both are decomposed to 1,000,000, so this place is equal.

Next, move one place to right on both numbers being compared. This number is zero for both numbers, so move on to the next number to the right. The first number decomposes to 20,000, while the second decomposes to 20,000. These numbers are also equal, so move one more place to the right. The first number decomposes to 3,000, as does the second number, so they are equal again. Moving one place to the right, the first number decomposes to 100, while the second number is zero. Since 100 is greater than zero, the first number is greater than the second. This is expressed using the greater than symbol:

1,023,100 > 1,023,000 because 1,023,100 is greater than 1,023,000 (Note that the "open mouth" of the symbol is nearest to 1,023,100).

Notice the > symbol in the above comparison. When values are the same, the equals sign (=) is used. However, when values are unequal, or an **inequality** exists, the relationship is denoted by various inequality symbols. These symbols describe in what way the values are unequal. A value could be greater than (>); less than (<); greater than or equal to (≥); or less than or equal to (≤) another value. The statement "five times a number added to forty is more than sixty-five" can be expressed as $5x + 40 > 65$. Common words and phrases that express inequalities are:

Symbol	Phrase
<	is under, is below, smaller than, beneath
>	is above, is over, bigger than, exceeds
≤	no more than, at most, maximum
≥	no less than, at least, minimum

Another way to compare whole numbers with many digits is to use place value. In each number to be compared, it is necessary to find the highest place value in which the numbers differ and to compare the value within that place value. For example, $4,523,345 < 4,532,456$ because of the values in the ten thousands place.

Rounding Whole Numbers

Rounding numbers changes the given number to a simpler and less accurate number than the exact given number. Rounding allows for easier calculations which estimate the results of using the exact given number. The accuracy of the estimate and ease of use depends on the place value to which the number is rounded. Rounding numbers consists of:

- Determining what place value the number is being rounded to
- Examining the digit to the right of the desired place value to decide whether to round up or keep the digit
- Replacing all digits to the right of the desired place value with zeros

To round 746,311 to the nearest ten thousands, the digit in the ten thousands place should be located first. In this case, this digit is 4 (7<u>4</u>6,311). Then, the digit to its right is examined. If this digit is 5 or greater, the number will be rounded up by increasing the digit in the desired place by one. If the digit to the right of the place value being rounded is 4 or less, the number will be kept the same. For the given example, the digit being examined is a 6, which means that the number will be rounded up by increasing the digit to the left by one. Therefore, the digit 4 is changed to a 5. Finally, to write the rounded number, any digits to the left of the place value being rounded remain the same and any to its right are replaced with zeros. For the given example, rounding 746,311 to the nearest ten thousand will produce 750,000. To round 746,311 to the nearest hundred, the digit to the right of the three in the hundreds place is examined to determine whether to round up or keep the same number. In this case, that digit is a one, so the number will be kept the same and any digits to its right will be replaced with zeros. The resulting rounded number is 746,300.

When rounding up, if the digit to be increased is a 9, the digit to its left is increased by 1 and the digit in the desired place value is changed to a zero. For example, the number 1,598 rounded to the nearest ten is 1,600. Another example shows the number 467,296 rounded to the nearest hundred thousand is 500,000 because the 6 in the ten-thousands place indicates that we need to round up. This changes the 4 in the hundred-thousands place to a 5.

Using Place Value Understanding and Properties of Operations to Perform Multi-Digit Arithmetic

Adding and Subtracting Multi-Digit Whole Numbers

Solving one-step and two-step problems with addition and subtraction requires knowledge of the vocabulary of operations, an understanding of place value, and identifying the relationship between addition and subtraction.

Gaining more of something related to addition, while taking something away relates to subtraction. Vocabulary words such as *total*, *more, less, left*, and *remain* are common when working with these problems. The + sign means plus. This shows that addition is happening. The − sign means minus. This shows that subtraction is happening. The symbols will be important when you write out equations.

Place value is utilized when performing operations. Operations are completed within each place value. Adding and subtracting multiples of tens can be introduced first. When adding or subtracting 10 or a multiple of 10 to a number, the nonzero value in the ones place does not change. Only the digits in the tens place need to be added or subtracted. For example, 98 − 60 = 38 is the result of keeping the 8 in the ones place and placing the solution of 9 − 6 = 3 in the tens place.

Adding and subtracting other numbers with more than one digit involves place value and rewriting numbers in expanded form. In the addition problem 256 + 261, 256 can be thought of as 200 + 50 + 6 or 2 hundreds, 5 tens, and 6 ones. 261 can be thought of as 200 + 60 + 1 or 2 hundreds, 6 tens, and 1 one. Adding the two numbers by place value results in 4 hundreds, 11 tens, and 7 ones. The 11 tens need to be regrouped as 1 hundred and 1 one. This leaves 5 hundreds, 1 ten, and 7 one, which is 517.

One method of subtraction involves a counting-up procedure. In the subtraction problem 476 − 241, adding 9 to 241 gives 250, adding 26 to 150 gives 276, and adding 200 to 276 gives 476. Therefore, the answer to the subtraction problem is $9 + 26 + 200 = 235$. The answer can be checked by adding 235 + 241 to make sure it equals 476. Also, the place value technique used within addition can be used rewriting each number in expanded form and then subtracting within each place value. Therefore, $400 + 70 + 6 - (200 + 40 + 1) = 200 + 30 + 5 = 235$. If one of the subtraction problems is not possible within a place value, the next largest place value must be regrouped. For instance, $262 - 71 = 200 + 60 + 2 - (70 + 1) = 100 + 160 + 2 - (70 - 2) = 100 + 90 + 1 = 191$.

Multiplying Multi-Digit Numbers

Multiplication involves adding together multiple copies of a number. It is indicated by an ×symbol or a number immediately outside of a parentheses, e.g. 5(8-2). The two numbers being multiplied together are called **factors,** and their result is called a **product**. For example, $9 \times 6 = 54$.

This can be shown alternatively by expansion of either the 9 or the 6:

$$9 \times 6 = 9 + 9 + 9 + 9 + 9 + 9 = 54$$

$$9 \times 6 = 6 + 6 + 6 + 6 + 6 + 6 + 6 + 6 + 6 = 54$$

Like addition, multiplication holds the commutative and associative properties:

$$115 = 23 \times 5 = 5 \times 23 = 115$$

$$84 = 3 \times (7 \times 4) = (3 \times 7) \times 4 = 84$$

Multiplication also follows the **distributive property**, which allows the multiplication to be distributed through parentheses. The formula for distribution is $a \times (b + c) = ab + ac$. This is clear after the examples:

$$45 = 5 \times 9 = 5(3 + 6) = (5 \times 3) + (5 \times 6) = 15 + 30 = 45$$

$$20 = 4 \times 5 = 4(10 - 5) = (4 \times 10) - (4 \times 5) = 40 - 20 = 20$$

One of the quickest methods to multiply larger numbers involves an **algorithm** to line up the products. For larger-number multiplication, how the numbers are lined up can ease the process. It is simplest to put the number with the most digits on top and the number with fewer digits on the bottom. If they have the same number of digits, select one for the top and one for the bottom. Line up the problem, and begin by multiplying the far-right column on the top and the far-right column on the bottom. If the answer to a column is more than 9, the one digit will be written below that column and the tens place digit will carry to the top of the next column to be added after those digits are multiplied. Write the answer below that column. Move to the next column to the left on the top, and multiply it by the same far right column on the bottom. Keep moving to the left one column at a time on the top number until the end.

Example: Multiply 37×8

Line up the numbers, placing the one with the most digits on top.

```
    3 7
x     8
------
```

Multiply the far-right column on the top with the far-right column on the bottom (7 x 8). Write the answer, 56, as below: The ones value, 6, gets recorded, the tens value, 5, is carried.

```
   +5
    3 7
X     8
------
2 9 6
```

Move to the next column left on the top number and multiply with the far-right bottom (3 x 8). Remember to add any carry over after multiplying: 3 x 8 = 24, 24 + 5 = 29. Since there are no more digits on top, write the entire number below.

```
   +5
    3 7
X     8
------
2 9 6
```

22

The solution is 296

Let's try multiplying a four-digit number by a one-digit number: $1,321 \times 3$.

First, line up the numbers on the far-right:

```
  1,321
×     3
```

Next, beginning on the far right and then moving one place at a time to the left, multiply the two numbers and write the answer below the line in the column of the top number. Continue the process until there are no more numbers to the left on top, as follows:

```
  1,321
×     3
      3
```

```
  1,321
×     3
     63
```

```
  1,321
×     3
    963
```

```
  1,321
×     3
  3,963
```

Some numbers will require a leading number to be carried up to the row to the left, as follows:

```
  1,621
×     3
      3
```

```
  1,621
×     3
     63
```

```
  1¹,621
×      3
     863
```
```
  1¹,621
×      3
   4,863
```

Notice that any carryovers are added to the sum of the two numbers being multiplied.

To use this algorithm for larger numbers, say a two-digit number being multiplied by another two-digit number, an additional step is necessary. If there is more than one column to the bottom number, move to the row below the first strand of answers, mark a zero in the far right column, and then begin the multiplication process again with the far right column on top and the second column from the right on the bottom. For each digit in the bottom number, there will be a row of answers, each padded with the respective number of zeros on the right. Finally, add up all of the answer rows for one total number.

For example, multiply 12 $\times$ 15.

Line up the numbers, as before.

```
   12
 × 15
```

Then begin multiplying on the far right, and move to the left.

```
   1¹2
 × 1 5
     0
```

```
   1¹2
 × 1 5
   6 0
```

Next, move one number to the left on the bottom row and begin the process again, while writing the products in the line under the first set of solutions.

The first row on the far right will automatically get a zero as a place holder, so the numbers will shift one column to the left of the original line up, as follows:

```
   12
 × 15
   6 0
   2 0
```

```
   12
 × 15
   6 0
 1 2 0
```

```
   12
 × 15
   6 0
 1 2 0
```

Now, add the two sets of solutions.

```
   12
 × 15
   6 0
 1 2 0
 1 8 0
```

Representing the Product of 2 Two-Digit Numbers

A simple way to represent the product of 2 two-digit numbers is through the use of arrays.

Consider this example of the product of 12 × 12 represented by an array:

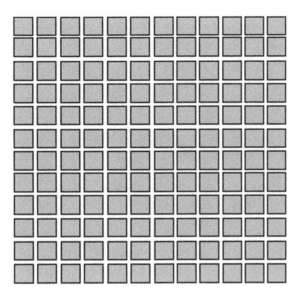

Since the resulting shape is square (12 tiles on each side) the total number of tiles represents a number that is called a **perfect square**, which is the product of a number multiplied by itself. An equation to describe this situation is $12 \times 12 = 144$.

144 is a perfect square.

Numbers that are not perfect squares can be represented in a similar fashion. For example, the product of 11 × 10 represented in an array looks like this:

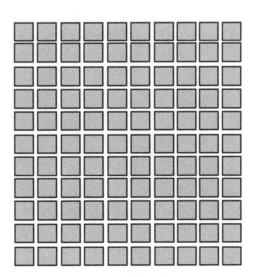

An equation to describe this situation is $11 \times 10 = 110$. This is not a square, so the total number of tiles is not a perfect square.

Any numbers can be represented using arrays, and then by an equation that describes the size of the array.

Another method of multiplication can be done with the use of an *area model*. An area model is a rectangle that is divided into rows and columns that match up to the number of place values within each number. For example, $29 \times 65 = 25 + 4$ and $66 = 60 + 5$. The products of those 4 numbers are found within the rectangle and then summed up to get the answer. The entire process is: $(60 \times 25) + (5 \times 25) + (60 \times 4) + (5 \times 4) = 1{,}500 + 240 + 125 + 20 = 1{,}885$. Here is the actual area model:

	25	4
60	60x25 1,500	60x4 240
5	5x25 125	5x4 20

$$
\begin{array}{r}
1{,}500 \\
240 \\
125 \\
+\quad 20 \\
\hline
1{,}885
\end{array}
$$

Dividing Whole Numbers with Multi-Digit Dividends

Division is based on dividing a given number into parts. The simplest problem involves dividing a number into equal parts. For example, a pack of 20 pencils is to be divided among 10 children. You would have to divide 20 by 10. In this example, each child would receive 2 pencils.

The symbol for division is $\div$ or $/$. The equation above is written as $20 \div 10 = 2$, or $20 / 10 = 2$. This means "20 divided by 10 is equal to 2." Division can be explained as the following: for any whole numbers a and b, where b is not equal to zero, $a \div b = c$ if and only if $a = b \times c$. This means, division can be thought of as a multiplication problem with a missing part. For instance, calculating $20 \div 10$ is the same as asking the following: "If there are 20 items in total with 10 in each group, how many are in each group?" Therefore, 20 is equal to ten times what value? This question is the same as asking, "If there are 20 items in total with 2 in each group, how many groups are there?" The answer to each question is 2.

In a division problem, a is known as the **dividend,** b is the **divisor**, and c is the **quotient**. Zero cannot be divided into parts. Therefore, for any nonzero whole number $a, 0 \div a = 0$. Also, division by zero is undefined. Dividing an amount into zero parts is not possible.

Division problems involve a total amount, a number of groups having the same amount, and a number of items within each group. The difference between multiplication and division is that the starting point is the total amount. It then gets divided into equal amounts.

For example, the equation is $15 \div 5 = 3$.

15 is the total number of items, which is being divided into 5 different groups. In order to do so, 3 items go into each group. Also, 5 and 3 are interchangeable. So, the 15 items could be divided into 3 groups of 5 items each. Therefore, different types of word problems can arise from this equation. For example, here are three types of problems:

- A boy needs 48 pieces of chalk. If there are 8 pieces in each box, how many boxes should he buy?

- A boy has 48 pieces of chalk. If each box has 6 pieces in it, how many boxes did he buy?

- A boy has partitioned all of his chalk into 8 piles, with 6 pieces in each pile. How many pieces does he have in total?

Each one of these questions involves the same equation. The third question can easily utilize the multiplication equation $8 \times 6 = ?$ instead of division. The answers are 6, 8, and 48.

Multiplication and division are inverse operations. So, multiplying by a number and then dividing by the same number results in the original number. For example, $8 \times 2 \div 2 = 8$ and $12 \div 4 \times 4 = 12$. Inverse operations are used to work backwards to solve problems. If a school's entire 4th grade was divided evenly into 3 classes each with 22 students, the inverse operation of multiplication is used to determine the total students in the grade ($22 \times 3 = 66$).

To calculate a division problem, a methodical algorithm can be followed, as modeled when calculating multiplication. For each portion, the number of times a divisor can evenly go into a dividend is tracked and collected to form the final solution, or **quotient**. The process begins where the dividend and the divisor meet on the left and progresses one spot to the left after any remainder is subtracted.

For example: $375 \div 4$.

$$4\overline{)375}$$ Set up the problem.

$$\begin{array}{r} 9 \\ 4\overline{)375} \\ 36 \\ \hline 1 \end{array}$$

Because 4 cannot divide into 3, add the next unit from the numerator, 7.
4 divides into 37, 9 times, so write the 9 above the 7.
Write the 36 under the 37 for subtraction; the remainder is 1 (1 is less than 4).

$$\begin{array}{r} 93 \\ 4\overline{)375} \\ 36\downarrow \\ \hline 15 \\ 12 \\ \hline 3 \end{array}$$

Drop down the next unit of the numerator, 5.
4 divides into 15, 3 times, so write the 3 above the 5.
Multiply 4 x 3.
Write the 12 under the 15 for subtraction; the remainder is 3 (3 is less than 4).

The solution is 93 remainder 3 OR $93\frac{3}{4}$ (the remainder can be written over the original denominator).

Representing the Quotient of Up to a Four-Digit Number Divided by a One-Digit Number
Dividing a number by a single digit or two digits can be turned into repeated subtraction problems. An area model can be used throughout the problem that represents multiples of the divisor. For example, the answer to $8580 \div 55$ can be found by subtracting 55 from 8580 one at a time and counting the total number of subtractions necessary.

However, a simpler process involves using larger multiples of 55. First, $100 \times 55 = 5,500$ is subtracted from 8,580, and 3,080 is leftover. Next, $50 \times 55 = 2,750$ is subtracted from 3,080 to obtain 380. $5 \times 55 = 275$ is subtracted from 330 to obtain 55, and finally, $1 \times 55 = 55$ is subtracted from 55 to obtain zero. Therefore, there is no remainder, and the answer is $100 + 50 + 5 + 1 = 156$.

Here is a picture of the area model and the repeated subtraction process:

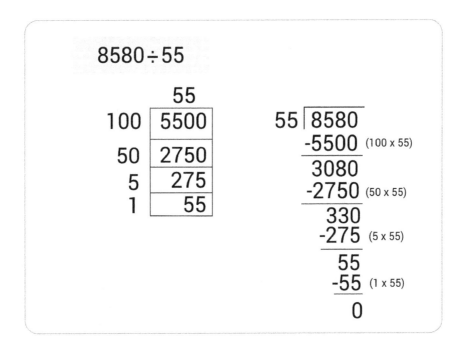

When given the task of trying to represent the quotient of up to a four-digit number divided by a one-digit number, another manageable method is utilizing an array.

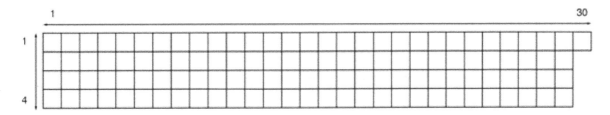

This array can be divided into three main sections. The first section is a 4 x 25 section that contains 100 pieces ($4 \times 25 = 100$). The second section is a 4 x 4 section that contains 16 pieces ($4 \times 4 = 16$). The final section is a single piece. These sections represent the division of the number 117 by 4. The sections are divided up into 4 portions of 25 plus 4 portions of 4, with one remaining portion. This leads to the following equation:

$$117 \div 4 = 25 + 4 \, R1$$

$$117 \div 4 = 29 \, R1$$

This shows how many times 4 can be divided into 117 and with any remainder (denoted by R).

Numbers and Operations: Fractions

Fraction Equivalence and Ordering

Identifying Equivalent Fractions

Like fractions, or **equivalent fractions**, represent two fractions that are made up of different numbers, but represent the same quantity. For example, the given fractions are $4/_8$ and $3/_6$. If a pie was cut into 8 pieces and 4 pieces were removed, half of the pie would remain. Also, if a pie was split into 6 pieces and 3 pieces were eaten, half of the pie would also remain. Therefore, both of the fractions represent half of a pie. These two fractions are referred to as like fractions. **Unlike fractions** are fractions that are different and cannot be thought of as representing equal quantities. When working with fractions in mathematical expressions, like fractions should be simplified. Both $4/_8$ and $3/_6$ can be simplified into $1/_2$.

Comparing fractions can be completed through the use of a number line. For example, if $\frac{3}{5}$ and $\frac{6}{10}$ need to be compared, each fraction should be plotted on a number line. To plot $\frac{3}{5}$, the area from 0 to 1 should be broken into 5 equal segments, and the fraction represents 3 of them. To plot $\frac{4}{10}$, the area from 0 to 1 should be broken into 10 equal segments and the fraction represents 6 of them.

It can be seen that $\frac{3}{5} = \frac{6}{10}$

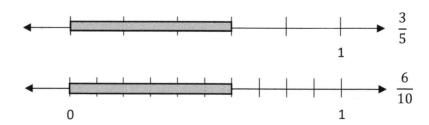

Like fractions are plotted at the same point on a number line.

Grouping objects or shading equivalent shapes are helpful in representing equivalent fractions. For example, if the fraction $\frac{1}{2}$ were to be represented by grouping objects totaling 6 and objects totaling 8, it can be shown through the following groupings:

$\frac{1}{2}$ of 6 is 3, or $\frac{3}{6}$:

$\frac{1}{2}$ of 8 is 4, or $\frac{4}{8}$:

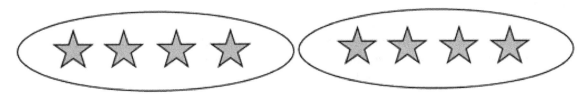

Comparing Fractions

To compare fractions with either the same **numerator** (top number) or same **denominator** (bottom number), it is easiest to visualize the fractions with a model.

For example, which is larger, $\frac{1}{3}$ or $\frac{1}{4}$? Both numbers have the same numerator, but a different denominator. In order to demonstrate the difference, shade the amounts on a pie chart split into the number of pieces represented by the denominator.

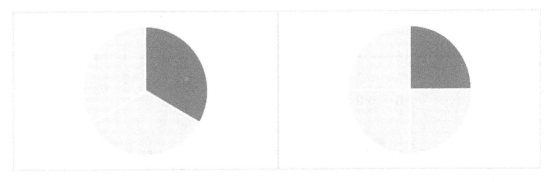

The first pie chart represents $\frac{1}{3}$, a larger shaded portion, and is therefore a larger fraction than the second pie chart representing $\frac{1}{4}$.

If two fractions have the same denominator (or are split into the same number of pieces), the fraction with the larger numerator is the larger fraction, as seen below in the comparison of $\frac{1}{3}$ and $\frac{2}{3}$:

$$\frac{1}{3} \qquad\qquad\qquad \frac{2}{3}$$

Remember, a **unit fraction** is one in which the numerator is 1 ($\frac{1}{2}, \frac{1}{3}, \frac{1}{8}, \frac{1}{20}$, etc.). The denominator indicates the number of equal pieces that the whole is divided into. The greater the number of pieces, the smaller each piece will be. Therefore, the greater the denominator of a unit fraction, the smaller it is in value. Unit fractions can also be compared by converting them to decimals. For example, $\frac{1}{2} = 0.5$, $\frac{1}{3} = 0.\overline{3}$, $\frac{1}{8} = 0.125$, $\frac{1}{20} = 0.05$, etc.

Comparing two fractions with different denominators can be difficult if attempting to guess at how much each represents. Using a number line, blocks, or just finding a common denominator with which to compare the two fractions makes this task easier.

For example, compare the fractions $\frac{3}{4}$ and $\frac{5}{8}$.

The number line method of comparison involves splitting one number line evenly into 4 sections, and the second number line evenly into 8 sections total, as follows:

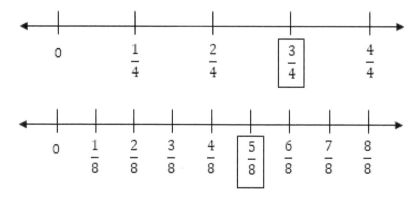

Here it can be observed that $\frac{3}{4}$ is greater than $\frac{5}{8}$, so the comparison is written as $\frac{3}{4} > \frac{5}{8}$.

This could also be shown by finding a common denominator for both fractions, so that they could be compared. First, list out factors of 4: 4, 8, 12, 16.

Then, list out factors of 8: 8, 16, 24.

Both share a common factor of 8, so they can be written in terms of 8 portions. In order for $\frac{3}{4}$ to be written in terms of 8, both the numerator and denominator must be multiplied by 2, thus forming the new fraction $\frac{6}{8}$. Now the two fractions can be compared.

Because both have the same denominator, the numerator will show the comparison.

$$\frac{6}{8} > \frac{5}{8}$$

Remember to use the inequality symbols when representing unequal comparisons:

Symbol	Phrase
<	is under, is below, smaller than, beneath
>	is above, is over, bigger than, exceeds
≤	no more than, at most, maximum
≥	no less than, at least, minimum

Building Fractions from Unit Fractions

Understanding a Fraction *a*/*b* with *a* > 1 as a Sum of Fractions 1/*b*

Adding and Subtracting Fractions

In order to add and subtraction fractions with unequal denominators, common denominators must be found. The **least common denominator** should be used as the common denominator, and this value is equal to the least common multiple of the denominators. For example, consider the following problem: $\frac{1}{5} + \frac{1}{2}$. Both fractions have different denominators, and the least common denominator is 10. Therefore, each fraction needs to be rewritten as an equivalent fraction with a denominator of 10 and then added. Pictures with fractions represented as pies can help us with this step. Consider the following picture:

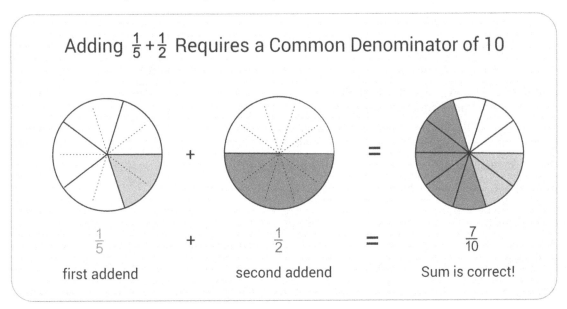

Adding $\frac{1}{5} + \frac{1}{2}$ Requires a Common Denominator of 10

$\frac{1}{5}$ + $\frac{1}{2}$ = $\frac{7}{10}$

first addend   second addend   Sum is correct!

The first pie shows that when it is split up into either 5 pieces or 10 pieces, $\frac{1}{5}$ and $\frac{2}{10}$ are equivalent. Similarly, the second pie shows that when it is split up into either 2 pieces or 10 pieces, $\frac{1}{2}$ and $\frac{5}{10}$ are equivalent. Therefore, the problem becomes $\frac{2}{10} + \frac{5}{10} = \frac{7}{10}$, which can be seen in the pie on the other side of the equals sign in the picture.

Composing and Decomposing a Fraction

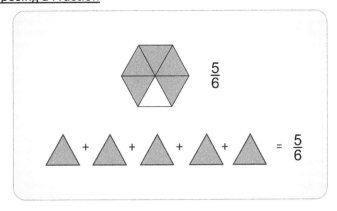

$\frac{5}{6}$

$\triangle + \triangle + \triangle + \triangle + \triangle = \frac{5}{6}$

Fractions can be broken apart into sums of fractions with the same denominator. For example, the fraction $\frac{5}{6}$ can be decomposed into sums of fractions with all denominators equal to 6 and the numerators adding to 5. The fraction $\frac{5}{6}$ is decomposed as:

$$\frac{3}{6}+\frac{2}{6}; \text{ or } \frac{2}{6}+\frac{2}{6}+\frac{1}{6}; \text{ or } \frac{3}{6}+\frac{1}{6}+\frac{1}{6}; \text{ or } \frac{1}{6}+\frac{1}{6}+\frac{1}{6}+\frac{2}{6}; \text{ or } \frac{1}{6}+\frac{1}{6}+\frac{1}{6}+\frac{1}{6}+\frac{1}{6}$$

A unit fraction is a fraction in which the numerator is 1. If decomposing a fraction into unit fractions, the sum will consist of a unit fraction added the number of times equal to the numerator. For example, $\frac{3}{4} = \frac{1}{4}+\frac{1}{4}+\frac{1}{4}$ (unit fractions $\frac{1}{4}$ added 3 times). Composing fractions is simply the opposite of decomposing. It is the process of adding fractions with the same denominators to produce a single fraction. For example, $\frac{3}{7}+\frac{2}{7} = \frac{5}{7}$ and $\frac{1}{5}+\frac{1}{5}+\frac{1}{5} = \frac{3}{5}$.

<u>Adding and Subtracting Mixed Numbers with Like Denominators,</u>

A **mixed number** is a whole number combined with a fraction. One example of a mixed number is $3\frac{2}{3}$. Notice that the numerator (the top number) is less than the denominator (the bottom number) in the fraction, so the fraction a called **proper fraction**. The entire mixed number can be converted to an **improper fraction**, which is a fraction in which the numerator is larger than the denominator. In order to do this, multiply the whole number by the denominator and then add the numerator. In our example of $3\frac{2}{3}$, this would result in $3 \times 3 + 2 = 11$. This result is then written over the original denominator. Therefore, the mixed number $3\frac{2}{3}$ is written as the improper fraction of $\frac{11}{3}$.

In order to add and subtract mixed numbers with like denominators, convert each mixed number to an improper fraction, add the numerators together, and write that result over the like denominator. For instance, consider $4\frac{1}{5} + 2\frac{2}{5}$. First, rewrite each mixed number as an improper fraction. Therefore, using the process just described, the problem becomes $\frac{21}{5} + \frac{12}{5}$. Then, add the numerators together over the common denominator, resulting in $\frac{33}{5}$. This result can be converted back into a mixed number by dividing 33 by 5, writing down the whole number, and then attaching the fraction, where the numerator is the remainder and the denominator is the denominator used throughout the entire problem. Therefore, we can write $\frac{33}{5}$ as $6\frac{3}{5}$.

<u>Solving Word Problems Involving Addition and Subtraction of Fractions</u>

If an area model is used to represent a fraction, a fraction can be shown visually by using parts of a whole area or region. Shapes such as rectangles and circles can represent the whole areas. If a mixed number is represented by an area model, the whole number portion is represented by that number of whole areas and the fraction portion is represented by a part of another whole. Here is an example of an area model used to represent the mixed number $2\frac{1}{2}$:

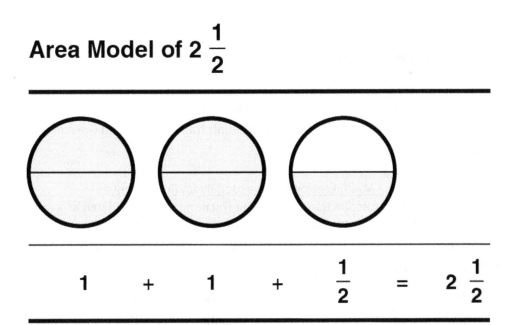

Note that 2 full circles are used to represent the whole number 2, and one half of the third circle is filled in to represent the $\frac{1}{2}$.

Such models can be used to solve word problems involving addition and subtraction of fractions with like denominators. For instance, let's say a classmate has selected 4 pencils out of a box of 10, and another classmate has selected 3 of the remaining pencils from the box. If you want to visualize the number of pencils each classmate took out of the total, and then calculate the fraction of the total they took together, we would utilize the following area model:

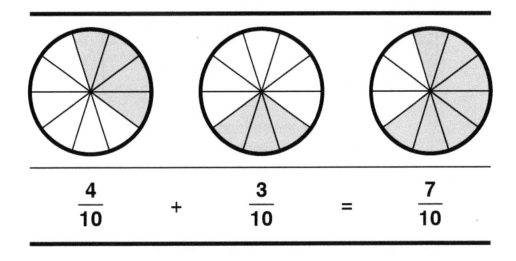

The first circle represents that 4 out of 10, or $\frac{4}{10}$, pencils were selected first, and the second circle represents that 3 out of 10, or $\frac{3}{10}$, were selected second. Adding them together, we have $\frac{4}{10} + \frac{3}{10} = \frac{7}{10}$ pencils. Note that 7 out of 10 pencils were removed from the box, which means 3 pencils (or $\frac{3}{10}$ of the box) remain.

Multiplying a Fraction by a Whole Number

Understanding a Fraction a/b as a Multiple of 1/b
A fraction a/b, like ¾, is a multiple of the unit fraction 1/b, or in this case, ¼. It is ¾ is 3 times the unit fraction. Written as an equation, $\frac{3}{4} = 3 \times \frac{1}{4}$. Recall that a **unit fraction** has a 1 in the numerator and an integer in the denominator, such as $\frac{1}{5}$ or $\frac{1}{8}$.

Solving Word Problems Involving Multiplication of a Fraction By a Whole Number
A whole number multiplied by a fraction that is not a unit fraction can be calculated as a multiple of a unit fraction with the same denominator. An example of a word problem where a whole number is multiplied by a fraction that is not a unit fraction is the following: Three friends each walk $\frac{3}{4}$ of a mile for charity. How far did they walk together?

First, we need to turn this work problem into a mathematical equation that can be solved. There are three people who each walk $\frac{3}{4}$ of a mile. Therefore, the equation is $3 \times \frac{3}{4}$. Three is the whole number and $\frac{3}{4}$ is the fraction and note that it is not a unit fraction. We can represent this process using pictures. First, draw three pictures that represent the fraction $\frac{3}{4}$ since we are multiplying $\frac{3}{4}$ by 3. Here, the fraction is represented by squares divided into four equal parts:

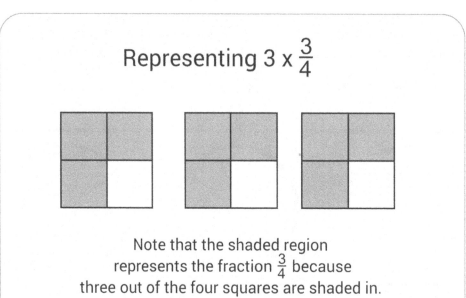

Representing 3 x $\frac{3}{4}$

Note that the shaded region represents the fraction $\frac{3}{4}$ because three out of the four squares are shaded in.

Note that the shaded region represents the fraction $\frac{3}{4}$ because three out of the four squares are shaded in.

Then, separate each square into three squares each containing one shaded in region as done in the following:

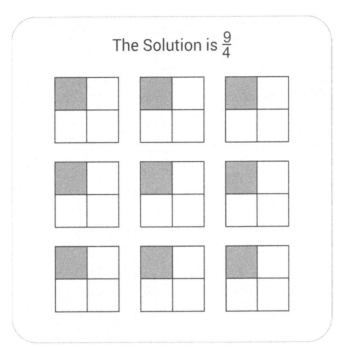

The Solution is $\frac{9}{4}$

Each square now represents the fraction $\frac{1}{4}$, and there are 9 of them, so the answer to our multiplication problem is $3 \times \frac{3}{4} = 9 \times \frac{1}{4} = \frac{9}{4}$. This means the three friends walked $\frac{9}{4}$ miles together. This improper fraction can be turned into the mixed number $2\frac{1}{4}$. This process can also be done using circles, where pieces of pie represent fractional parts of the whole.

Understanding Decimal Notation for Fractions, and Comparing Decimal Fractions

Decimals and fractions are two ways that can be used to represent positive numbers less than one. Formally, a **decimal** is a number that has a dot within the number. For example, 3.45 is a decimal, and the dot is called a **decimal point**. The number to the left of the decimal point is in the ones place. The number to the right of the decimal point represents the portion of the number less than one. The first number to the right of the decimal point is the tenths place, and one tenth represents $^1/_{10}$, just like a dime. The next place is the hundredths place, and it represents $^1/_{100}$, just like a penny. This idea is continued to the right in the hundredths, thousandths, and ten thousandths places. Each place value to the right is ten times smaller than the one to its left.

A number less than one contains only digits in some decimal places. For example, 0.53 is less than one. A **mixed number** is a number greater than one that also contains digits in some decimal places. For example, 3.43 is a mixed number. Adding a zero to the right of a decimal does not change the value of the number. For example, 2.75 is the same as 2.750. However, 2.75 is the more accepted representation of the number. Also, zeros are usually placed in the ones column in any value less than one. For example, 0.65 is the same as .65, but 0.65 is more widely used.

In order to read or write a decimal, the decimal point is ignored. The number is read as a whole number, and then the place value unit is stated in which the last digit falls. For example, 0.089 is read as *eighty-nine thousandths*, and 0.1345 is read as *one thousand, three hundred forty-five ten thousandths*. In mixed numbers, the word "and" is used to represent the decimal point. For example, 2.56 is read as *two and fifty-six hundredths*.

Counting money—specifically, quantities less than one dollar—is a good method to better understand values less than one. Recall that the four common coins (pennies, nickels, dimes, and quarters) are worth portions of one dollar:

- Quarter = $0.25

- Dime = $0.10

- Nickel = $0.05

- Penny = $0.01

Notice the decimal place. It indicates that the values of each of these coins is equal to just a portion of one dollar ($1.00). These coins, also called **change,** can be combined in multiple ways in order to equal different amounts of money, and are often in story problems that are applicable to real-world situations.

For example, Smith wants to purchase a number of items at the candy counter. Prices are as follows:

Gum = $0.05
Chocolate = $0.20
Licorice = $0.10
Cherry sours = $0.15

Smith wants to buy 1 piece of chocolate, 4 pieces of gum, 1 piece of licorice, and 1 cherry sour. Would he be able to use the following coins, and if so, what are possible combinations Smith could use?

To begin, add the total value of the coins:

2 quarters + 2 dimes + 1 nickel

$$(2 \times \$0.25) + (2 \times \$0.10) + \$0.05 = \$0.75$$

Then calculate the total cost of the items Smith wants to purchase:

1 chocolate + 4 pieces of gum + 1 licorice + 1 cherry sour

$$\$0.20 + (4 \times \$0.05) + \$0.10 + \$0.15 = \$0.65$$

Smith has enough money to purchase all of the items, and there is only one combination that would provide the correct amount of $0.65: 2 quarters, 1 dime, and 1 nickel.

How much change would Smith receive if he paid with a one-dollar bill, and what would the possible combinations be from the assortment pictured above?

$$\$1.00 - \$0.65 = \$0.35$$

The only coin combination for this change is 1 quarter and 1 dime.

Consider another example: If a student had three quarters and a dime and wanted to purchase a cookie at lunch for 50 cents, how much change would she receive? The answer is found by first calculating the sum of the change as 85 cents and then subtracting 50 cents to get 35 cents.

Expressing a Fraction with Denominator 10 as an Equivalent Fraction with Denominator 100

If two fractions have different denominators, equivalent fractions must be used to add or subtract them. If you are given two fractions with different denominators and asked to add them together, it would be wrong to add the numerators together without adjusting the original fractions. The fractions must first be converted into equivalent fractions that have common denominators (the numbers on the bottom of each fraction must be the same). If one fraction has a denominator of 10 and the other has a denominator of 100, the fraction with a denominator of 10 must be converted to an equivalent fraction with a denominator of 100. For example, consider $\frac{2}{10} + \frac{27}{100}$. They cannot be added together as written now because they do not have like denominators. The first fraction needs to be converted to an equivalent fraction with a denominator of 100. In order to do this, because the denominator is being multiplied by 10 to result in 100, the numerator must be multiplied by 10 as well. Therefore, the equivalent fraction is $\frac{2 \times 10}{10 \times 10} = \frac{20}{100}$. The problem now becomes $\frac{20}{100} + \frac{27}{100}$. Both fractions now have like denominators and the numerators can be added together and then written over the common denominator to complete the addition process. The result is $\frac{20}{100} + \frac{27}{100} = \frac{47}{100}$.

Using Decimal Notation for Fractions with Denominators 10 or 100

Fractions with denominators of 10 or 100 can easily be converted to decimal notation. This is because the base-10 system we use for place value has a specific place for tenths and hundredths. The tenths place is one place to the right of the decimal point, so one tenth or $\frac{1}{10}$ is written as .1 (you can also include the zero to the left of the decimal point without changing the value: 0.1). Similarly, the fraction $\frac{4}{10}$, which is read as "four tenths," can be written .4.

The same concept applies for fractions with a denominator of 100 because these fractions represent parts of one hundred, or hundredths, just like the place value two spaces to the right of the decimal point. For example, $\frac{7}{100}$ is written .07.

Basically, because a fraction can be thought of as a division problem (the numerator (or top number) is divided by the denominator (or bottom number), to convert a fraction to a decimal, the numerator is divided by the denominator. For example, $\frac{3}{8}$ can be converted to a decimal by dividing 3 by 8 ($\frac{3}{8} = 0.375$). Therefore, the fraction $\frac{3}{8}$ is equivalent to the decimal 0.375.

To convert a decimal to a fraction, the decimal point is dropped and the value is written as the numerator. The denominator is the place value farthest to the right with a digit other than zero. For example, to convert .48 to a fraction, the numerator is 48 and the denominator is 100 (the digit 8 is in the hundredths place). Therefore, $.48 = \frac{48}{100}$. Fractions should be written in the simplest form, or reduced. To reduce a fraction, the numerator and denominator are divided by the largest common factor. In the previous example, 48 and 100 are both divisible by 4. Dividing the numerator and denominator by 4 results in a reduced fraction of $\frac{12}{25}$.

To represent fractions and decimals as distances beginning at zero on a number line, it's helpful to relate the fraction to a real-world application. For example, a charity walk covers $\frac{3}{10}$ of a mile. How could this distance be represented on a number line?

First, divide the number line into tenths, as follows:

If each division on the number line represents one-tenth of one, or $\frac{1}{10}$, then representing the distance of the charity walk, $\frac{3}{10}$, would cover 3 of those divisions and look as follows:

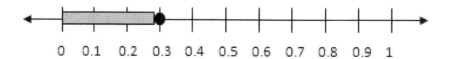

So, the fraction $\frac{3}{10}$ is represented by covering from 0 to 0.3 (or 3 sections) on the number line.

Comparing and Ordering Decimals

To compare decimals and order them by their value, utilize a method similar to that of ordering large numbers.

The main difference is where the comparison will start. Assuming that any numbers to left of the decimal point are equal, the next numbers to be compared are those immediately to the right of the decimal point. If those are equal, then move on to compare the values in the next decimal place to the right.

For example:

Which number is greater, 12.35 or 12.38?

Check that the values to the left of the decimal point are equal:

$$12 = 12$$

Next, compare the values of the decimal place to the right of the decimal:

$$12.3 = 12.3$$

Those are also equal in value.

Finally, compare the value of the numbers in the next decimal place to the right on both numbers:

12.3**5** and 12.3**8**

Here the 5 is less than the 8, so the final way to express this inequality is:

12.35 < 12.38

Notice the < symbol in the above comparison. When values are the same, the equals sign (=) is used. However, when values are unequal, or an **inequality** exists, the relationship is denoted by various inequality symbols. These symbols describe in what way the values are unequal. A value could be greater than (>); less than (<); greater than or equal to (≥); or less than or equal to (≤) another value. The statement "five times a number added to forty is more than sixty-five" can be expressed as $5x + 40 > 65$. Common words and phrases that express inequalities are:

Symbol	Phrase
<	is under, is below, smaller than, beneath
>	is above, is over, bigger than, exceeds
≤	no more than, at most, maximum
≥	no less than, at least, minimum

Comparing decimals is regularly exemplified with money because the "cents" portion of money ends in the hundredths place. When paying for gasoline or meals in restaurants, and even in bank accounts, if enough errors are made when calculating numbers to the hundredths place, they can add up to dollars and larger amounts of money over time.

Number lines can also be used to compare decimals. Tick marks can be placed within two whole numbers on the number line that represent tenths, hundredths, etc. Each number being compared can then be plotted. The leftmost value on the number line is the largest.

Determining Decimals on a Number Line
To precisely understand a number being represented on a number line, the first step is to identify how the number line is divided up. When utilizing a number line to represent decimal portions of numbers, it is helpful to label the divisions, or insert additional divisions, as needed.

For example, what number, to the nearest hundredths place is marked by the point on the following number line?

First, figure out how the number line is divided up. In this case, it has ten sections, so it is divided into tenths. To use this number line with the divisions, label the divisions as follows:

Because the dot is placed equally between 0.4 and 0.5, it is at 0.45.

What number, to the nearest tenths place, is marked by the point on the following number line?

First, determine what the number line is divided up into and mark it on the line.

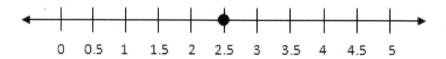

This number line is divided up into half of the whole numbers it represents, or 0.5 increments. The division and labeling of the number line assists in easily reading the dot as marking 2.5.

Measurement and Data

Solving Problems Involving Measurement and Conversions

The Relative Sizes of Measurement Units and Converting Units

Measurement is how an object's length, width, height, weight, and so on, are quantified. Measurement is related to counting, but it is a more refined process.

The United States customary system and the metric system each consist of distinct units to measure lengths and volume of liquids. The U.S. customary units for length, from smallest to largest, are: inch (in), foot (ft), yard (yd), and mile (mi). The metric units for length, from smallest to largest, are: millimeter (mm), centimeter (cm), decimeter (dm), meter (m), and kilometer (km). The relative size of each unit of length is shown below.

U.S. Customary	Metric	Conversion
12in = 1ft	10mm = 1cm	1in = 254cm
36in = 3ft = 1yd	10cm = 1dm(decimeter)	1m ≈ 3.28ft ≈ 1.09yd
5,280ft = 1,760yd = 1mi	100cm = 10dm = 1m	1mi ≈ 1.6km
	1000m = 1km	

The U.S. customary units for volume of liquids, from smallest to largest, are: fluid ounces (fl oz), cup (c), pint (pt), quart (qt), and gallon (gal). The metric units for volume of liquids, from smallest to largest, are: milliliter (mL), centiliter (cL), deciliter (dL), liter (L), and kiloliter (kL). The relative size of each unit of liquid volume is shown below.

U.S. Customary	Metric	Conversion
8fl oz = 1c	10mL = 1cL	1pt ≈ 0.473L
2c = 1pt	10cL = 1dL	1L ≈ 1.057qt
4c = 2pt = 1qt	1,000mL = 100cL = 10dL = 1L	1gal ≈ 3,785L
4qt = 1gal	1,000L = 1kL	

The U.S. customary system measures weight (how strongly Earth is pulling on an object) in the following units, from least to greatest: ounce (oz), pound (lb), and ton. The metric system measures mass (the quantity of matter within an object) in the following units, from least to greatest: milligram (mg), centigram (cg), gram (g), kilogram (kg), and metric ton (MT).

The relative sizes of each unit of weight and mass are shown below.

U.S. Measures of Weight	Metric Measures of Mass
16oz = 1lb	10mg = 1cg
2,000lb = 1 ton	100cg = 1g
	1,000g = 1kg
	1,000kg = 1MT

Note that weight and mass DO NOT measure the same thing.

Time is measured in the following units, from shortest to longest: second (sec), minute (min), hour (h), day (d), week (wk), month (mo), year (yr), decade, century, millennium. The relative sizes of each unit of time is shown below.

- 60sec = 1min
- 60min = 1h
- 24hr = 1d
- 7d = 1wk
- 52wk = 1yr
- 12mo = 1yr
- 10yr = 1 decade
- 100yrs = 1 century
- 1,000yrs = 1 millennium

Converting Measurements

When working with different systems of measurement, conversion from one unit to another may be necessary. The conversion rate must be known to convert units. One method for converting units is to write and solve a proportion. The arrangement of values in a proportion is extremely important. Suppose that your doctor measures your height in inches to be 51 inches and you want to know if you will be tall enough to ride the big roller coasters, which require a height of 4 feet. Therefore, the problem requires converting your height in inches to feet.

Looking at the conversion chart, it can be seen that 1 foot = 12 inches. A proportion can be set up using this conversion and x representing the missing value.

The proportion can be written in any of the following ways:

$$\frac{1}{12} = \frac{x}{51} \left(\frac{feet\ for\ conversion}{inches\ for\ conversion} = \frac{unknown\ feet}{inches\ given} \right)$$

$$\frac{12}{1} = \frac{51}{x} \left(\frac{inches\ for\ conversion}{feet\ for\ conversion} = \frac{inches\ given}{unknown\ feet} \right)$$

$$\frac{1}{x} = \frac{12}{51} \left(\frac{feet\ for\ conversion}{unknown\ feet} = \frac{inches\ for\ conversion}{inches\ given} \right)$$

$$\frac{x}{1} = \frac{51}{12} \left(\frac{unknown\ feet}{feet\ for\ conversion} = \frac{inches\ given}{inches\ for\ conversion} \right)$$

To solve a proportion, the ratios are cross-multiplied and the resulting equation is solved. When cross-multiplying, all four proportions above will produce the same equation: $(12)(x) = (51)(1) \rightarrow 12x = 51$. Dividing by 12 to isolate the variable x, the result is $x = 4.25$. The variable x represented the unknown number of feet. Therefore, the conclusion is that 51 inches converts (is equal) to 4.25 feet, meaning you are tall enough to ride the roller coasters!

Note that while there are four options presented for how to correctly set up the proportion, not every arrangement is correct. It would be incorrect, for example to set up the following:

$$\frac{1}{12} = \frac{51}{x} \left(\frac{feet\ for\ conversion}{inches\ for\ conversion} = \frac{inches\ given}{unknown\ feet} \right)$$

This is because when cross-multiplying, it would be seen that inches would be multiplied by inches and feet by feet instead of inches with feet. Therefore, the units would not cancel out and the proportion is incorrect.

Let's try another problem. Suppose that a problem requires converting 20 fluid ounces to cups. To do so, a proportion can be written using the conversion rate of 8fl oz = 1c with x representing the missing value. Again, the proportion can be written in any of the following ways:

$$\frac{1}{8} = \frac{x}{20} \left(\frac{c \; for \; conversion}{fl \; oz \; for \; conversion} = \frac{unknown \; c}{fl \; oz \; given} \right)$$

$$\frac{8}{1} = \frac{20}{x} \left(\frac{fl \; oz \; for \; conversion}{c \; for \; conversion} = \frac{fl \; oz \; given}{unknown \; c} \right)$$

$$\frac{1}{x} = \frac{8}{20} \left(\frac{c \; for \; conversion}{unknown \; c} = \frac{fl \; oz \; for \; conversion}{fl \; oz \; given} \right)$$

$$\frac{x}{1} = \frac{20}{8} \left(\frac{unknown \; c}{c \; for \; conversion} = \frac{fl \; oz \; given}{fl \; oz \; for \; conversion} \right)$$

To solve the proportion, the ratios are cross-multiplied and the resulting equation is solved. When cross-multiplying, all four proportions above will produce the same equation: $(8)(x) = (20)(1) \rightarrow 8x = 20$. Dividing by 8 to isolate the variable x, the result is $x = 2.5$. The variable x represented the unknown number of cups. Therefore, the conclusion is that 20 fluid ounces converts (is equal) to 2.5 cups.

Sometimes converting units requires writing and solving more than one proportion. Suppose an exam question asks to determine how many hours are in 2 weeks. Without knowing the conversion rate between hours and weeks, this can be determined knowing the conversion rates between weeks and days, and between days and hours. First, weeks are converted to days, then days are converted to hours. To convert from weeks to days, the following proportion can be written:

$$\frac{7}{1} = \frac{x}{2} \left(\frac{days \; conversion}{weeks \; conversion} = \frac{days \; unknown}{weeks \; given} \right)$$

Cross-multiplying produces: $(7)(2) = (x)(1) \rightarrow 14 = x$. Therefore, 2 weeks is equal to 14 days. Next, a proportion is written to convert 14 days to hours:

$$\frac{24}{1} = \frac{x}{14} \left(\frac{conversion \; hours}{conversion \; days} = \frac{unknown \; hours}{given \; days} \right)$$

Cross-multiplying produces: $(24)(14) = (x)(1) \rightarrow 336 = x$. Therefore, the answer is that there are 336 hours in 2 weeks.

Solving Problems Involving Measurements

Problems that involve measurements of length, time, volume, etc. are generally dependent upon understanding how to manipulate between various units of measurement, as well as understanding their equivalencies.

Determining Solutions to Problems Involving Time

Time is measured in units such as *seconds, minutes, hours, days,* and *years*. For example, there are 60 seconds in a minute, 60 minutes in each hour, and 24 hours in a day.

When dealing with problems involving elapsed time, break the problem down into workable parts. For example, suppose the length of time between 1:15pm and 3:45pm must be determined. From 1:15pm to 2:00pm is 45 minutes (knowing there are 60 minutes in an hour). From 2:00pm to 3:00pm is 1 hour. From 3:00pm to 3:45pm is 45 minutes. The total elapsed time is 45 minutes plus 1 hour plus 45 minutes. This sum produces 1 hour and 90 minutes. 90 minutes is over an hour, so this is converted to 1 hour (60 minutes) and 30 minutes. The total elapsed time can now be expressed as 2 hours and 30 minutes.

To illustrate time intervals, a clock face can show solutions.

For example, Ani needs to complete all of her chores by 1:50 p.m. If she begins her chores at 1:00 p.m., can she finish the following? Vacuuming (15 minutes), dusting (10 minutes), replacing light bulbs (5 minutes), and degreasing the garage floor (25 minutes).

A blank clock face is useful in illustrating the time lapse necessary for all of Ani's tasks.

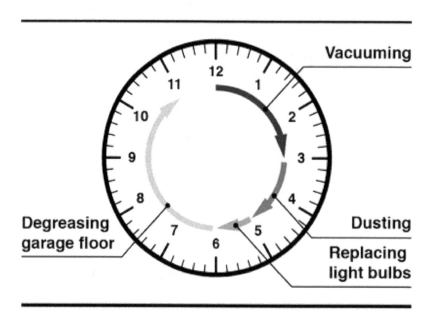

It is easy to see that the chores will span beyond 50 minutes after the hour, so no, Ani could not complete the chores in the given time frame.

When dealing with problems involving elapsed time, breaking the problem down into workable parts is helpful. For example, suppose the length of time between 1:15pm and 3:45pm must be determined. From 1:15pm to 2:00pm is 45 minutes (knowing there are 60 minutes in an hour). From 2:00pm to 3:00pm is 1 hour. From 3:00pm to 3:45pm is 45 minutes. The total elapsed time is 45 minutes plus 1 hour plus 45 minutes. This sum produces 1 hour and 90 minutes. 90 minutes is over an hour, so this is converted to 1 hour (60 minutes) and 30 minutes. The total elapsed time can now be expressed as 2 hours and 30 minutes.

Determining Solutions to Problems Involving Money

Let's consider a problem involving change. Gwen wants to purchase a number of items at the school supply store. Prices are as follows:

Erasers = $0.05
Highlighters = $0.20
Pencils = $0.10
Pens = $0.15

Gwen wants to buy 1 highlighter, 5 erasers, 2 pencils, and 3 pens. Would she be able to use the following coins, and if so, what are possible combinations Gwen could use?

To begin, add the total value of the coins:

$$2 \text{ quarters} + 5 \text{ dimes} + 2 \text{ nickel}$$

$$(2 \times \$0.25) + (5 \times \$0.10) + (2 \times \$0.05) = \$1.10$$

Then calculate the total cost of the items Gwen wants to purchase:

Erasers = $0.05
Highlighters = $0.20
Pencils = $0.10
Pens = $0.15

Gwen wants to buy 1 highlighter, 5 erasers, 2 pencils, and 3 pens.

$$1 \text{ highlighter} + 5 \text{ erasers} + 2 \text{ pencils} + 2 \text{ pens}$$

$$\$0.20 + (5 \times \$0.05) + (2 \times \$0.10) + (2 \times \$0.15)$$

$$\$0.20 + \$0.25 + \$0.20 + \$0.30 = \$0.95$$

Gwen has enough money to purchase all of the items, and there is only one combination that would provide the correct amount of $0.95: 2 quarters, 4 dimes, and 1 nickel.

Here is another example of a problem involving money that is a bit more difficult:

A store is having a spring sale, where everything is 70% off. You have $45.00 to spend. A jacket is regularly priced at $80.00. Do you have enough to buy the jacket and a pair of gloves, regularly priced at $20.00?

There are two ways to approach this.

Method 1:

Set up the equations to find the sale prices: the original price minus the amount discounted.
$80.00 - ($80.00 (0.70)) = sale cost of the jacket.
$20.00 – ($20.00 (0.70)) = sale cost of the gloves.
Solve for the sale cost.
$24.00 = sale cost of the jacket.
$6.00 = sale cost of the gloves.
Determine if you have enough money for both.
$24.00 + $6.00 = total sale cost.
$30.00 is less than $45.00, so you can afford to purchase both.

Method 2:

Determine the percent of the original price that you will pay.
100% – 70% = 30%
Set up the equations to find the sale prices.
$80.00 (0.30) = cost of the jacket.
$20.00 (0.30) = cost of the gloves.
Solve.
$24.00 = cost of the jacket.
$6.00 = cost of the gloves.
Determine if you have enough money for both.
$24.00 + $6.00 = total sale cost.
$30.00 is less than $45.00, so you can afford to purchase both.

<u>Determining Solutions to Problems Involving Length</u>
The length of an object can be measured using standard tools such as rulers, yard sticks, meter sticks, and measuring tapes. The following image depicts a yardstick:

Choosing the right tool to perform the measurement requires determining whether United States customary units or metric units are desired, and having a grasp of the approximate length of each unit and the approximate length of each tool. The measurement can still be performed by trial and error without the knowledge of the approximate size of the tool.

For example, to determine the length of a room in feet, a United States customary unit, various tools can be used for this task. These include a ruler (typically 12 inches/1 foot long), a yardstick (3 feet/1 yard long), or a tape measure displaying feet (typically either 25 feet or 50 feet). Because the length of a room is much larger than the length of a ruler or a yardstick, a tape measure should be used to perform the measurement.

When the correct measuring tool is selected, the measurement is performed by first placing the tool directly above or below the object (if making a horizontal measurement) or directly next to the object (if making a vertical measurement). The next step is aligning the tool so that one end of the object is at the mark for zero units, then recording the unit of the mark at the other end of the object. To give the length of a paperclip in metric units, a ruler displaying centimeters is aligned with one end of the paper clip to the mark for zero centimeters.

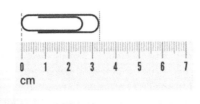

Directly down from the other end of the paperclip is the mark that measures its length. In this case, that mark is two small dashes past the 3 centimeter mark. Each small dash is 1 millimeter (or .1 centimeters). Therefore, the length of the paper clip is 3.2 centimeters.

To compare the lengths of objects, each length must be expressed in the same unit. If possible, the objects should be measured with the same tool or with tools utilizing the same units. For example, a ruler and a yardstick can both measure length in inches. If the lengths of the objects are expressed in different units, these different units must be converted to the same unit before comparing them. If two lengths are expressed in the same unit, the lengths may be compared by subtracting the smaller value from the larger value. For example, suppose the lengths of two gardens are to be compared. Garden A has a length of 4 feet, and garden B has a length of 2 yards. 2 yards is converted to 6 feet so that the measurements have similar units. Then, the smaller length (4 feet) is subtracted from the larger length (6ft): 6ft – 4ft = 2ft. Therefore, garden B is 2 feet larger than garden A.

Identifying and utilizing the proper units for the scenario requires knowing how to apply the conversion rates for length. For example, given a scenario that requires subtracting 8 inches from $2\frac{1}{2}$ feet, both values should first be expressed in the same unit (they could be expressed $\frac{2}{3}$ft & $2\frac{1}{2}$ft, or 8in and 30in). The desired unit for the answer may also require converting back to another unit.

Consider the following scenario that involves length: A parking area along the river is only wide enough to fit one row of cars and is $\frac{1}{2}$ kilometers long. The average space needed per car is 5 meters. How many cars can be parked along the river? First, all measurements should be converted to similar units: $\frac{1}{2}$km = 500m. The operation(s) needed should be identified. Because the problem asks for the number of cars, the total space should be divided by the space per car. 500 meters divided by 5 meters per car yields a total of 100 cars. Written as an expression, the meters unit cancels and the cars unit is left: $\frac{500m}{5m/car}$ the same as $500m \times \frac{1\,car}{5m}$ yields 100 cars.

Determining Solutions to Problems Involving Volume
Volume is how much space something occupies. That "something" can be liquid or solid. Essentially, volume is a measurement of capacity. Whereas area is calculated by counting squares within a two-dimensional object, volume is calculated by counting cubes within a three-dimensional object. It is a measure of the space a figure occupies. Volume is measured using cubic units, such as cubic inches, feet, centimeters, or kilometers. Centimeter cubes can be utilized in the classrooms in order to promote understanding of volume.

49

For instance, if 10 cubes were placed along the length of a rectangle, with 8 cubes placed along its width, and the remaining area was filled in with cubes, there would 80 cubes in total, which would equal a volume of 80 cubic centimeters. Its area would equal 80 square centimeters. If that shape was doubled so that its height consists of two cube lengths, there would be 160 cubes, and its volume would be 160 cubic centimeters. Adding another level of cubes would mean that there would be $3 \times 80 = 240$ cubes. This idea shows that volume is calculated by multiplying area times height. The actual formula for volume of a three-dimensional rectangular solid is $V = l \times w \times h$, where l represents length, w represents width, and h represents height. Volume can also be thought of as area of the base times the height. The base in this case would be the entire rectangle formed by l and w. Here is an example of a rectangular solid with labeled sides:

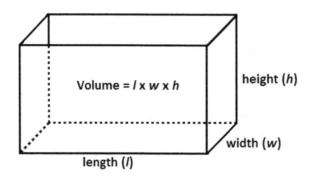

A **cube** is a special type of rectangular solid in which its length, width, and height are the same. If this length is s, then the formula for the volume of a cube is $V = s \times s \times s$.

The U.S. customary units for volume of liquids, from smallest to largest, are: fluid ounces (fl oz), cup (c), pint (pt), quart (qt), and gallon (gal). The metric units for volume of liquids, from smallest to largest, are: milliliter (mL), centiliter (cL), deciliter (dL), liter (L), and kiloliter (kL.).

A problem involving liquid volume would be: If Mart needed 2 quarts of liquid for a recipe and only has a measuring cup, how could he measure out 2 quarts?

The solution would involve Mart measuring out 2 quarts by filling the cup 8 times.

Determining Solutions to Problems Involving Mass
The metric system measures **mass**, which is the quantity of matter within an object. Mass and weight do not measure the same thing. **Weight** is affected by gravity, and deals with how strongly Earth is pulling on an object.

The following is an example of a problem involving mass:

A piggy bank contains 12 dollars' worth of nickels. The mass of a nickel is 5 grams, and the empty piggy bank has a mass of 1050 grams. What is the total mass of the full piggy bank?

A dollar contains 20 nickels. Therefore, if there are 12 dollars' worth of nickels, there are $12 \times 20 = 240$ nickels. The mass of each nickel is 5 grams. Therefore, the mass of the nickels is $240 \times 5 = 1,200$ grams. Adding in the mass of the empty piggy bank, the mass of the filled bank 2,250 grams.

Applying the Area and Perimeter Formulas for Rectangles in Real World and Mathematical Problems

Perimeter is the length of all its sides. The perimeter of a given closed sided figure would be found by first measuring the length of each side and then calculating the sum of all sides. A rectangle consists of two sides called the length (*l*), which have equal measures, and two sides called the width (*w*), which have equal measures. Therefore, the perimeter (*P*) of a rectangle can be expressed as $P = l + l + w + w$. This can be simplified to produce the following formula to find the perimeter of a rectangle: $P = 2l + 2w$ or $P = 2(l + w)$.

The perimeter of a square is measured by adding together all of the sides. Since a square has four equal sides, its perimeter can be calculated by multiplying the length of one side by 4. Thus, the formula is $P = 4 \times s$, where *s* equals one side. For example, the following square has side lengths of 5 meters:

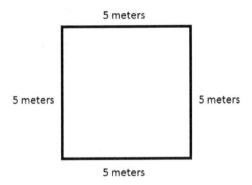

5 meters

5 meters   5 meters

5 meters

The perimeter is 20 meters because 4 times 5 is 20.

Like a square, a rectangle's perimeter is measured by adding together all of the sides. But as the sides are unequal, the formula is different. A rectangle has equal values for its lengths (long sides) and equal values for its widths (short sides), so the perimeter formula for a rectangle is:

$$P = l + l + w + w = 2l + 2w$$

l equals length
w equals width

Consider the following problem:

The total perimeter of a rectangular garden is 36 m. If the length of each side is 12 m, what is the width?

The formula for the perimeter of a rectangle is P=2L+2W, where P is the perimeter, L is the length, and W is the width. The first step is to substitute all of the data into the formula:

$$36 = 2(12) + 2W$$

Simplify by multiplying 2x12:

$$36 = 24 + 2W$$

Simplifying this further by subtracting 24 on each side, which gives:

$$36-24 = 24-24+2W$$

$$12= 2W$$

Divide by 2:

$$6 = W$$

The width is 6 m. Remember to test this answer by substituting this value into the original formula:

$$36 = 2(12) + 2(6)$$

The **area** of a two-dimensional figure refers to the number of square units needed to cover the interior region of the figure. This concept is similar to wallpaper covering the flat surface of a wall. For example, if a rectangle has an area of 10 square centimeters (written $10cm^2$), it will take 10 squares, each with sides one centimeter in length, to cover the interior region of the rectangle. Note that area is measured in square units such as: square centimeters or cm^2; square feet or ft^2; square yards or yd^2; square miles or mi^2.

The area of a rectangle is found by multiplying its length, l, times its width, w. Therefore, the formula for area is $A = l \times w$. An equivalent expression is found by using the term base, b, instead of length, to represent the horizontal side of the shape. In this case, the formula is $A = b \times h$. This same formula can be used for all parallelograms.

Here is a visualization of a rectangle with its labeled sides:

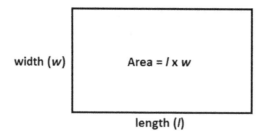

Using the unit square, the area of a rectangle can be calculated by tiling it with unit squares and then counting the squares. As an example, let's consider a rectangle with a length of 4 units and a width of 3 units. The goal is to know how many square units there are in total (the area of the rectangle). Three rows of four squares gives $4 + 4 + 4 = 12$. Also, three times four squares gives $3 \times 4 = 12$. Therefore, for any whole numbers a and b, where a is not equal to zero, $a \times b = b + b + \cdots b$, where b is added a times. Also, $a \times b$ can be thought of as the number of units in a rectangular block consisting of a rows and b columns.

For example, 3×7 is equal to the number of squares in the following rectangle:

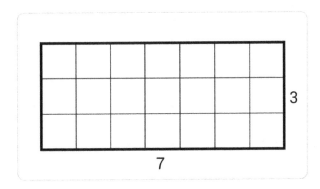

The answer is 21, and there are 21 squares in the rectangle. The area of the rectangle, therefore, is also 21 square units.

The concept of area can be related to the operations of multiplication and addition. Multiplication can be thought of as repeated addition. To model the relationship between area and these operations, organizing objects into arrays or groups of equal numbers for combining is helpful.

For example, a jewelry store's sales are represented by ring boxes in the following diagram. How many total sales were there?

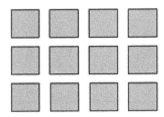

The array above of three rows by four columns (or 3 x 4) shows a total of 12 boxes. Notice that the array can be seen as comprising either three rows of four boxes each, or four columns of three boxes each. This array is akin to a model to calculate the area of a rectangle that is 3 x 4.

Here is another way to look at the jewelry boxes, but with four rows and three columns (a 4 x 3 array).

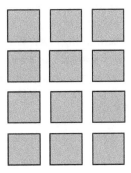

These two diagrams demonstrate the **commutative property of multiplication**, the idea that numbers can be multiplied in any order.

The area of a square is the length of a side squared, and the area of a rectangle is length multiplied by the width. For example, if the length of the square is 7 centimeters, then the area is 49 square centimeters. The formula for this example is $A = s^2 = 7^2 = 49$ square centimeters. An example is if the rectangle has a length of 6 inches and a width of 7 inches, then the area is 42 square inches:

$$A = lw = 6(7) = 42 \text{ square inches}$$

Representing and Interpreting Data

Making and Using Line Plots to Display a Data Set of Measurements in Fractions of a Unit

A **line plot** is a display of data that is plotted along a number line that shows **frequency**, or how often the same data exists. The frequency is drawn using either an x or a dot. The following is a line plot drawn that represents the data containing the numbers 0, 1, 2, 3, 4, 5. This data could represent the number of siblings students in a class have. Note that because there are 6 x's at the 2 on the plot, there must be 6 occurrences of 2 in the data. That means six students have 2 siblings.

Number of Siblings of Students in Mr. Schwartz's Class

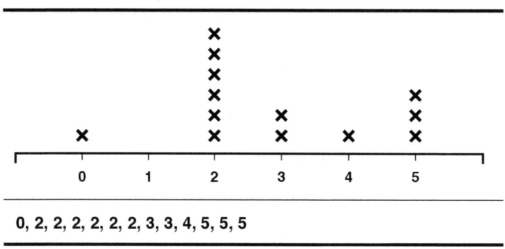

0, 2, 2, 2, 2, 2, 2, 3, 3, 4, 5, 5, 5

A line plot can be used to display a data set of measurements in fractions of units as well. The reason why this type of plot is necessary would be to organize data of different measurements in a real-world situation. Not every measurement will be a whole number. For instance, let's say you collected 15 leaves from the backyard and measured them to the nearest quarter inch. Your measurements were 3 quarter-inch leaves, 6 half-inch leaves, 5 three quarter-inch leaves, and 1 one-inch leaves.

The line plot that represents this data is shown here:

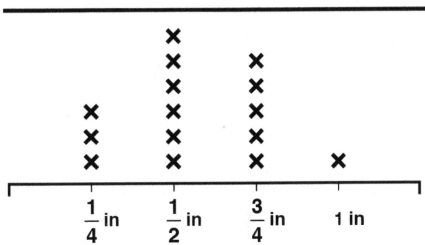

Leaf Lengths

Note that the number line shows the four different measurement possibilities and the x's represent the frequency, or number of times, that each measurement appeared. This line plot can be used to answer questions about the data set. For instance, the tallest height at the half-inch mark shows that this was the most frequent found measurement. Also, the chart could be used to find the difference in lengths between the longest and shortest leave found, which is called the **range** of the measurements. The longest leaf was 1 inch and the shortest leaf was a quarter-inch, so the difference is found through subtraction, with a result of $\frac{3}{4}$ inch.

Geometric measurement: Understanding Concepts of Angle and Measuring Angles

Recognizing Angles and Understanding Angle Measurements

<u>Measuring Angles and Understanding One-Degree Angles and Multiples</u>
An **angle** is created through the intersection of two line segments, known as **rays**. The intersection point of the two rays is known as a **vertex**. The amount of turn between the two rays within a single angle is known as the **measure of the angle,** which is usually measured in either degrees or radians. The symbol for degrees is °. The measurement is based on a circle that is comprised of 360 degrees. A full turn around any circle is 360 degrees, a half turn around any circle is 180 degrees, and a quarter turn around any circle is 90 degrees. A 90-degree angle is referred to as a **right angle**. This looks like a hard corner, which is like that in a square or rectangle. If an angle only turns $\frac{1}{360}$ of a turn within a circle, the angle is known as a one-degree angle. The idea of a one-degree angle can be used to measure any angle by counting the number of one-degree turns within the angle.

Here is a picture of a few angles measured within a circle:

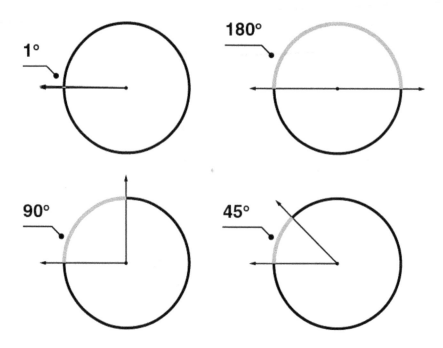

The two rays or an angle can be called sides of the angle. The angle below has a vertex at point *B* and the sides consist of ray *BA* and ray *BC*. An angle can be named in three ways:

1. Using the vertex and a point from each side, with the vertex letter in the middle.
2. Using only the vertex. This can only be used if it is the only angle with that vertex.
3. Using a number that is written inside the angle.

The angle below can be written $\angle ABC$ (read angle *ABC*), $\angle CBA$, $\angle B$, or $\angle 1$.

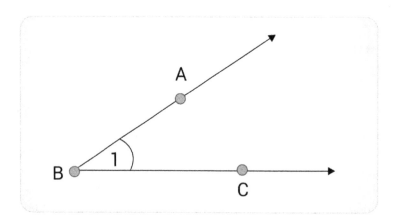

An angle divides a plane, or flat surface, into three parts: the angle itself, the interior (inside) of the angle, and the exterior (outside) of the angle. The figure below shows point *M* on the interior of the angle and point *N* on the exterior of the angle.

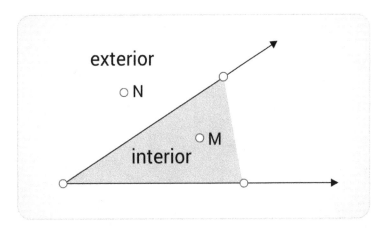

The Measure of an Angle with *n* Degrees

An angle does not have to be a perfect full, half, or quarter turn of a circle. For instance, here are some various angle measures:

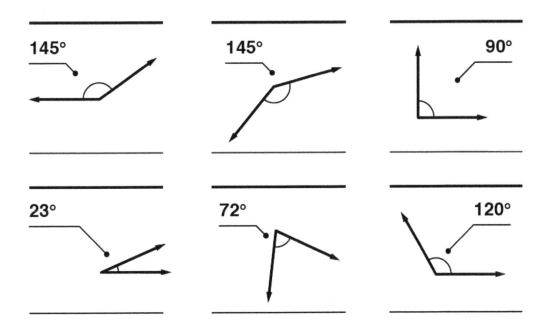

In general, an angle that turns through *n* one-degree angles has an angle measure of *n* degrees. In the picture, the degree symbol is °. Note the 23-degree angle has an angle measure of 23 degrees and is composed of 23 one-degree angles added together.

A **protractor** can be used to measure angles. In order to utilize this tool, align one of the rays with the 0 on the protractor and align the vertex with the center of the tool. Then, move your eyes out to the other ray within the angle and the corresponding number on the protractor through which that ray passes.

This number is the measure of the angle. Here is an example of a 40-degree angle measured by using a protractor:

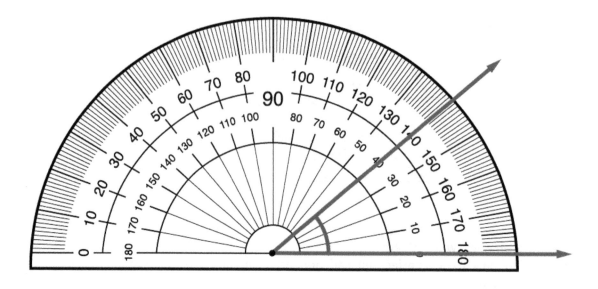

Note that depending on the direction of the angle, one of the rays can start on either the left-hand 0 side or the right-hand 0 side.

Measuring Angles in Whole-Number Degrees Using a Protractor
Angles can be measured in units called degrees, with the symbol °. The degree measure of an angle is between 0° and 180°, is a measure of rotation, and can be obtained by using a protractor.

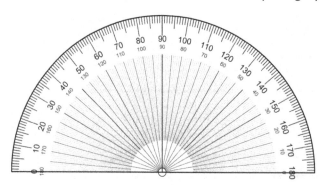

To use a protractor to measure an angle, the vertex, or corner, of the angle goes in the midpoint of the protractor, in that small circle along the bottom straight edge. Then one line of the angle is lined up along the bottom edge, toward the 0° indicator. Then, the degrees are read wherever the other line of the angle crosses.

In the example below, it can be seen that the angle measures about 30°:

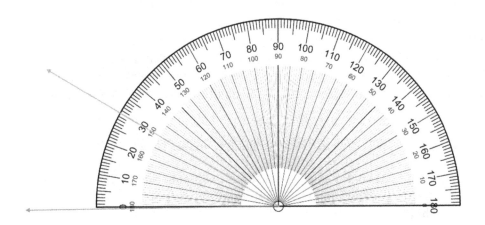

A straight angle (or simply a line) measures exactly 180°. A right angle's sides meet at the vertex to create a square corner. A right angle measures exactly 90° and is typically indicated by a box drawn in the interior of the angle. An acute angle has an interior that is narrower than a right angle. The measure of an acute angle is any value less than 90° and greater than 0°. For example, 89.9°, 47°, 12°, and 1°. An obtuse angle has an interior that is wider than a right angle. The measure of an obtuse angle is any value greater than 90° but less than 180°. For example, 90.1°, 110°, 150°, and 179.9°. Any two angles that sum up to 90 degrees are known as **complementary angles**.

- Acute angles: Less than 90°
- Obtuse angles: Greater than 90°
- Right angles: 90°
- Straight angles: 180°

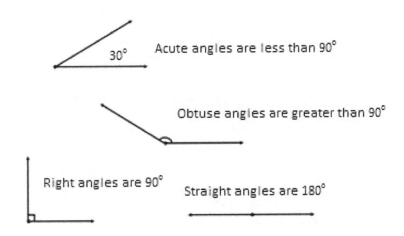

These definitions can be used to help sketch an approximation of the angle. If, for example, an 80° angle is to be sketched, a right angle can be used as the starting image in one's mind. Then instead of having exactly perpendicular lines in a hard corner, the angle can be made a little sharper (more acute).

59

When a more precise drawing of an angle is needed, protractors are used because they can help us draw angles with a given measure. The little hole at the midpoint along the bottom flat diameter of the circle is where the dot for the vertex of the angle should be drawn. Then a straight line is drawn using the straight edge of the protractor. This will be the ray of the angle that runs through the 0° mark on the protractor (again, along the straight part along the bottom). Then, look around the outer rim of the protractor until you find the desired number for the angle you want to draw. For example, if tasked with drawing a 40° angle, look for the number 40. Then draw a small point along the outer edge of the protractor where the 40° mark is (or whatever mark you need for your specific angle). Then, use the straight edge to draw a ray that runs from the vertex through the point you just made. You now should have drawn the appropriate angle.

Recognizing Angle Measures as Additive and Solving Addition and Subtraction Problems to Find Unknown Angles

To determine angle measures for adjacent angles, angles that share a common side and vertex, at least one of the angles must be known. Other information that is necessary to determine such measures include that there are 90° in a right angle, and there are 180° in a straight line. Therefore, if two adjacent angles form a right angle, they will add up to 90°, and if two adjacent angles form a straight line, they add up to 180°.

If the measurement of one of the adjacent angles is known, the other can be found by subtracting the known angle from the total number of degrees.

For example, given the following situation, if angle *a* measures 55°, find the measure of unknown angle *b*:

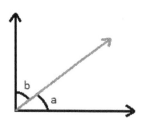

To solve this simply subtract the known angle measure from 90°.

$$90° - 55° = 35°$$

The measure of *b* = 35°.

Given the following situation, if angle 1 measures 45°, find the measure of the unknown angle 2:

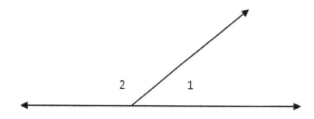

To solve this, simply subtract the known angle measure from 180°.

$$180° - 45° = 135°$$

The measure of angle 2 = 135°.

In the case that more than two angles are given, use the same method of subtracting the known angles from the total measure.

For example, given the following situation, if angle *y* = 40⁰, and angle *z* = 25⁰, find unknown angle *x*.

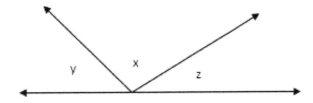

Subtract the known angles from 180⁰.

$$180° - 40° - 25° = 115°$$

The measure of angle *x* = 115⁰.

Geometry

Drawing and Identifying Lines and Angles, and Classifying Shapes by Properties of Their Lines and Angles

Identifying Points, Lines, Line Segments, Rays, and Angles

The basic unit of geometry is a point. A **point** represents an exact location on a plane, or flat surface. The position of a point is indicated with a dot and usually named with a single uppercase letter, such as point *A* or point *T*. A point is a place, not a thing, and therefore has no dimensions or size. A set of points that lies on the same line is called **collinear**. A set of points that lies on the same plane is called **coplanar**.

The image above displays point *A*, point *B*, and point *C*.

A **line** is as series of points that extends in both directions without ending. It consists of an infinite number of points and is drawn with arrows on both ends to indicate it extends infinitely. Lines can be named by two points on the line or with a single, cursive, lower case letter. The two lines below could be named line *AB* or line *BA* or $\overleftrightarrow{AB}$ or $\overleftrightarrow{BA}$; and line *m*.

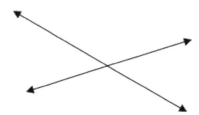

Two lines are considered parallel to each other if, while extending infinitely, they will never intersect (or meet). **Parallel** lines point in the same direction and are always the same distance apart. Two lines are considered **perpendicular** if they intersect to form right angles. Right angles are 90°. Typically, a small box is drawn at the intersection point to indicate the right angle.

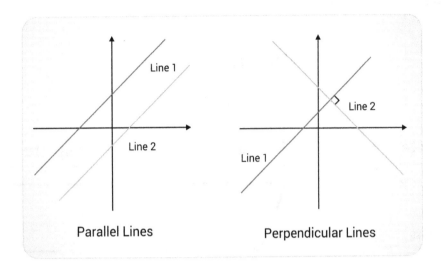

Line 1 is parallel to line 2 in the left image and is written as line 1 || line 2. Line 1 is perpendicular to line 2 in the right image and is written as line 1 ⊥ line 2.

A **ray** has a specific starting point and extends in one direction without ending. The endpoint of a ray is its starting point. Rays are named using the endpoint first, and any other point on the ray. The following ray can be named ray AB and written $\overrightarrow{AB}$.

An **angle** can be visualized as a corner. It is defined as the formation of two rays connecting at a vertex that extend indefinitely.

A **line segment** has specific starting and ending points. A line segment consists of two endpoints and all the points in between. Line segments are named by the two endpoints. The example below is named segment KL or segment LK, written $\overline{KL}$ or $\overline{LK}$.

Classifying Two-Dimensional Figures Based on the Presence or Absence of Parallel or Perpendicular Lines, or the Presence or Absence of Angles of a Specified Size

Lines that are **parallel** never intersect and run in the same direction. **Perpendicular** lines intersect at a 90-degree angles. Shapes can be classified whether they have these types of lines. For instance, a **parallelogram** has four sides in which each pair of opposite sides is parallel. Squares and rectangles meet the requirements of parallelograms. If a four-sided figure does not have pairs of parallel sides, it is known as a general quadrilateral. If it has one pair of parallel sides, it is a **trapezoid**. Here is a picture that shows different quadrilaterals:

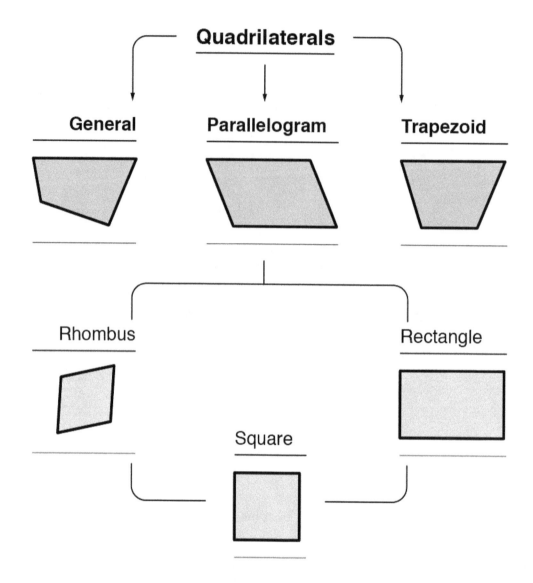

A right angle is 90 degrees. An acute angle has a measure less than 90 degrees, and an obtuse angle has a measure greater than 90 degrees. Triangles can be classified based on what type of angles they contain. A right triangle has one right angle, an acute triangle has three angles that have measure less than 90 degrees, and an obtuse angle has one angle that is greater than 90 degrees. Here is a picture that contains all three types of angles in their corresponding triangles:

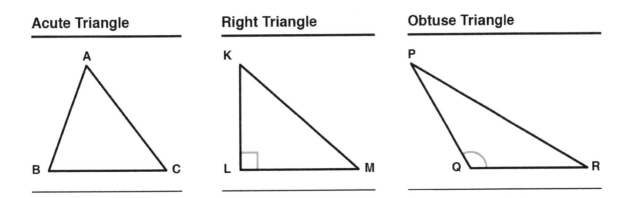

Four-sided shapes can also be classified based on angle measure. For instance, a square and a rectangle both have four 90-degree angles. All other quadrilaterals have mixture of acute, obtuse, and right angles. Squares and rectangles are classified as equiangular polygons. An **equiangular polygon** is a polygon that is composed of equal angles.

A triangle with three 45-degree angles is another example of an equiangular polygons. A **pentagon** that is equiangular is composed of five 108-degree angles, a **hexagon** that is equiangular is composed of six 120-degree angles, and an **octagon** that is equiangular is composed of eight 135-degree angles.

Right Triangles
Triangles can be further classified by their sides and angles. A triangle with its largest angle measuring 90° is a **right triangle**. A 90° angle, also called a **right angle,** is formed from two perpendicular lines. It looks like a hard corner, like that in a square. The little square draw into the angle is the symbol used to denote that that angle is indeed a right angle. Any time that symbol is used, it denotes the measure of the angle is 90°. Below is a picture of a right angle, and below that, a right triangle.

A right angle:

Here is a right triangle, which is a triangle that contains a right angle:

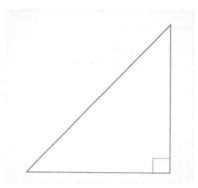

A triangle with the largest angle less than 90° is an acute triangle. A triangle with the largest angle greater than 90° is an obtuse triangle.

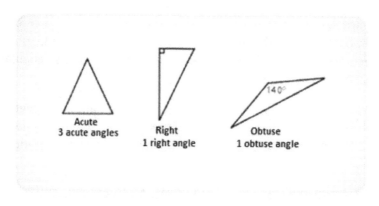

A triangle consisting of two equal sides and two equal angles is an isosceles triangle. A triangle with three equal sides and three equal angles is an equilateral triangle. A triangle with no equal sides or angles is a scalene triangle.

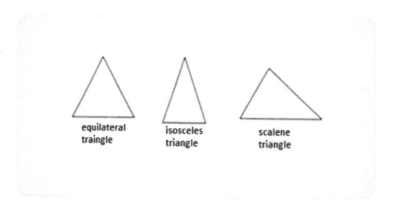

Recognizing Lines of Symmetry

Symmetry is another concept in geometry. If a two-dimensional shape can be folded along a straight line and the halves line up exactly, the figure is **symmetric**. The line is known as a **line of symmetry**. Circles, squares, and rectangles are examples of symmetric shapes.

Below is an example of a pentagon with a line of symmetry drawn.

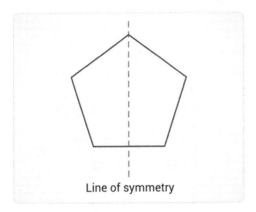

Line of symmetry

If a line cannot be drawn anywhere on the object to flip the figure onto itself, the object is **asymmetrical**. An example of a shape with no line of symmetry would be a scalene triangle.

If an object can be rotated about its center to any degree smaller than 360, and it lies directly on top of itself, the object is said to have **rotational symmetry**. An example of this type of symmetry is shown below. The pentagon has an order of 5.

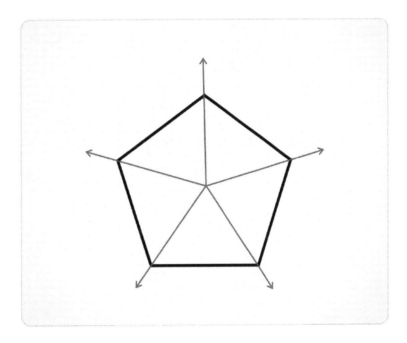

The rotational symmetry lines in the figure above can be used to find the angles formed at the center of the pentagon. Knowing that all of the angles together form a full circle, at 360 degrees, the figure can be split into 5 angles equally. By dividing the 360° by 5, each angle is 72°.

Given the length of one side of the figure, the perimeter of the pentagon can also be found using rotational symmetry. If one side length was 3 cm, that side length can be rotated onto each other side length four times. This would give a total of 5 side lengths equal to 3 cm. To find the perimeter, or distance around the figure, multiply 3 by 5. The perimeter of the figure would be 15 cm.

If a line cannot be drawn anywhere on the object to flip the figure onto itself or rotated less than or equal to 180 degrees to lay on top of itself, the object is asymmetrical. Examples of these types of figures are shown below.

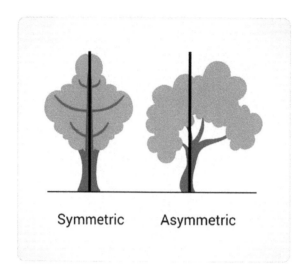

Symmetric   Asymmetric

No line of symmetry

Practice Questions

1. How will the number 84,789 be written if rounded to the nearest hundred?
    a. 84,800
    b. 84,700
    c. 84,780
    d. 85,000

2. One digit in the following number is in **bold** and the other is underlined: 3<u>6</u>,**6**01.
Which of the following statement about the underlined digit is true?
    a. Its value is $\frac{1}{10}$ the value of the bold digit.
    b. Its value is 10 times the value of the bold digit.
    c. Its value is 100 times the value of the bold digit.
    d. Its value is 60 times the value of the bold digit.

3. Angie wants to shade $\frac{3}{5}$ of this strip. Which is the correct representation of $\frac{3}{5}$?

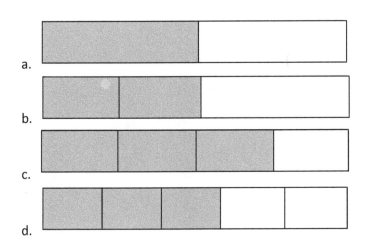

    a.
    b.
    c.
    d.

4. Ming would like to share his collection of 16 baseball cards with his three friends. He has decided that he will divide the collection equally among himself and his friends. Which of the following shows the correct grouping of Ming's cards?

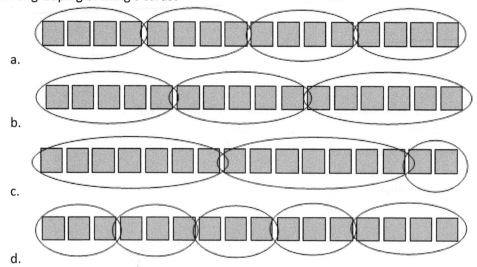

a.

b.

c.

d.

5. Which fractions are equivalent, or would fill the same portion on a number line?

   a. $\frac{2}{4}$ and $\frac{3}{8}$

   b. $\frac{1}{2}$ and $\frac{4}{8}$

   c. $\frac{3}{6}$ and $\frac{3}{5}$

   d. $\frac{2}{4}$ and $\frac{5}{8}$

6. Which fraction represents the greatest part of the whole?

   a. $\frac{1}{4}$

   b. $\frac{1}{3}$

   c. $\frac{1}{5}$

   d. $\frac{1}{2}$

7. Which of the following expressions is equivalent to $\frac{8}{7}$

   a. $\frac{1}{7}+\frac{1}{7}+\frac{1}{7}+\frac{1}{7}+\frac{1}{7}+\frac{1}{7}+\frac{1}{7}+\frac{1}{7}$

   b. $\frac{1}{8}+\frac{1}{8}+\frac{1}{8}+\frac{1}{8}+\frac{1}{8}+\frac{1}{8}+\frac{1}{8}$

   c. $\frac{1}{7}+\frac{8}{1}$

   d. $\frac{7}{8}+7$

8. Chris walks $\frac{4}{7}$ of a mile to school and Tina walks $\frac{5}{9}$ of a mile. Which student covers more distance on the walk to school?

    a. Chris, because $\frac{4}{7} > \frac{5}{9}$

    b. Chris, because $\frac{4}{7} < \frac{5}{9}$

    c. Tina, because $\frac{5}{9} > \frac{4}{7}$

    d. Tina, because $\frac{5}{9} < \frac{4}{7}$

9. A closet is filled with red, blue, and green shirts. If $\frac{1}{3}$ of the shirts are green and $\frac{2}{5}$ are red, what fraction of the shirts are blue?

    a. $\frac{4}{15}$

    b. $\frac{1}{5}$

    c. $\frac{7}{15}$

    d. $\frac{1}{2}$

10. Shawna buys $2\frac{1}{2}$ gallons of paint. If she uses $\frac{1}{3}$ of it on the first day, how much does she have left?

    a. $1\frac{5}{6}$ gallons

    b. $1\frac{1}{2}$ gallons

    c. $1\frac{2}{3}$ gallons

    d. 2 gallons

The following stem-and-leaf plot shows plant growth in cm for a group of tomato plants.

Stem	Leaf
2	0 2 3 6 8 8 9
3	2 6 7 7
4	7 9
5	4 6 9

11. What is the range of measurements for the tomato plants' growth?
   a. 29 cm
   b. 37 cm
   c. 39 cm
   d. 59 cm

12. How many plants grew more than 35 cm?
   a. 4 plants
   b. 5 plants
   c. 8 plants
   d. 9 plants

13. Which of the following is equivalent to the value of the digit 3 in the number 792.134?
   a. 3×10

   b. 3×100

   c. $\dfrac{3}{10}$

   d. $\dfrac{3}{100}$

14. A rectangle was formed out of pipe cleaner. Its length was $\frac{1}{2}$ feet and its width was $\frac{11}{2}$ inches. What is its area in square inches?
   a. $\frac{11}{4}$ inch^2
   b. $\frac{11}{2}$ inch^2
   c. 22 inch^2
   d. 33 inch^2

15. Which of the following statements is true about the two lines below?

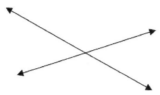

    a. The two lines are parallel but not perpendicular.
    b. The two lines are perpendicular but not parallel.
    c. The two lines are both parallel and perpendicular.
    d. The two lines are neither parallel nor perpendicular.

16. Which of the following numbers is greater than (>) 220,058?
    a. 220,158
    b. 202,058
    c. 220,008
    d. 217,058

17. What is the value, to the nearest tenths place, of the point indicated on the following number line?

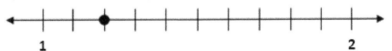

    a. 0.2
    b. 1.4
    c. 1.2
    d. 2.2

18. What two fractions add up to $\frac{7}{6}$?
    a. $\frac{2}{3} + \frac{5}{3}$
    b. $\frac{1}{5} + \frac{6}{5}$
    c. $\frac{1}{6} + \frac{6}{6}$
    d. $\frac{1}{2} + \frac{6}{4}$

19. Which represents the number 0.65 on a number line?

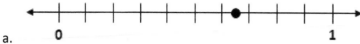

a.

b.

c.

d.

20. What equation, involving the addition of two fractions, is represented on the following number line?

a. $\frac{4}{5} + \frac{3}{5} = \frac{7}{5}$

b. $\frac{4}{5} + \frac{7}{5} = \frac{7}{5}$

c. $\frac{3}{5} + \frac{3}{5} = \frac{6}{5}$

d. $\frac{4}{5} + 1\frac{3}{5} = \frac{7}{5}$

21. Which of the following shows a line of symmetry?

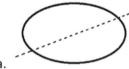

a.

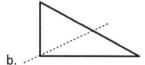

b.

c.

d.

22. The perimeter of a 6-sided polygon is 56 cm. The length of three of the sides are 9 cm each. The length of two other sides are 8 cm each. What is the length of the missing side?

    a. 11 cm
    b. 12 cm
    c. 13 cm
    d. 10 cm

23. Which four-sided shape is always a rectangle?

    a. Rhombus
    b. Square
    c. Parallelogram
    d. Quadrilateral

24. If Amanda can eat two times as many mini cupcakes as Marty, what would the missing values be for the following input-output table?

Input (number of cupcakes eaten by Marty)	Output (number of cupcakes eaten by Amanda)
1	2
3	
5	10
7	
9	18

    a. 6, 10
    b. 3, 11
    c. 6, 14
    d. 4, 12

25. Which of the following is not a parallelogram?

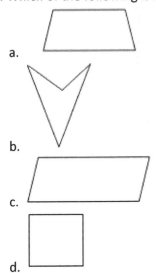

    a.

    b.

    c.

    d.

26. What is the measure of angle 2 be in the diagram below?

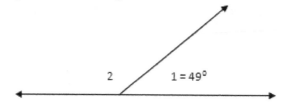

2   1 = 49⁰

    a. 131⁰
    b. 41⁰
    c. 311⁰
    d. 49⁰

27. Mo needs to buy enough material to cover the walls around the stage for a theater performance. If he needs 79 feet of wall covering, what is the minimum number of yards of material he should purchase if the material is sold only by whole yards?
    a. 23 yards
    b. 25 yards
    c. 26 yards
    d. 27 yards

28. The following table shows the temperature readings in Ohio during the month of January. How many more times was the temperature between 36-38 degrees than between 28-32 degrees?

Maximum Temperatures in degrees	Tally marks	Frequency
20 - 22	I	1
22 - 24	ЈН II	7
24 - 26	ЈН	5
26 - 28	ЈН IIII	9
28 - 30	ЈН ЈН	10

    a. 9 times
    b. 5 times
    c. 4 times
    d. 10 times

29. A rectangle was formed out of pipe cleaner. Its length was 3 in, and its width was 8 inches. What is its area in square inches?

    a. 22 inch^2
    b. 11 inch^2
    c. 32 inch^2
    d. 24 inch^2

30. Taylor is buying things at the bake sale at his sister's basketball game. Here is the price list:

- Brownies: $0.50
- Cookies: $.20
- Cupcakes: $.75
- Lemon Squares: $.60
- Milk: $35

He buys 1 brownie, 2 cookies, 1 cupcake, 2 lemon squares, and 1 container of milk. He gives the cashier the following money:

How much change should he receive?

    a. $0.20
    b. $0.25
    c. $0.40
    d. $0.75

31. Kassidy drove for 3 hours at a speed of 60 miles per hour. Using the distance formula, $d = r \times t$ ($distance = rate \times time$), how far did Kassidy travel?

    a. 20 miles
    b. 180 miles
    c. 65 miles
    d. 120 miles

32. It costs Shea $12 to produce 3 necklaces. If he can sell each necklace for $20, how much profit would he make if he sold 60 necklaces?

    a. $240
    b. $360
    c. $960
    d. $1200

33. The phone bill is calculated each month using the equation $c = 50g + 75$. The cost of the phone bill per month is represented by c, and g represents the gigabytes of data used that month. What is the value and interpretation of the slope of this equation?
    a. 75 dollars per day
    b. 75 gigabytes per day
    c. 50 dollars per day
    d. 50 dollars per gigabyte

34. Mom's car drove 72 miles in 90 minutes. How fast did she drive in feet per second?
    a. 0.8 feet per second
    b. 48.9 feet per second
    c. 0.009 feet per second
    d. 70. 4 feet per second

35. Which of the following are correct labels for the chart below?

Input	Calculation (Input × 3)	Output
1	1 × 3	3
2	2 × 3	6
3	3 × 3	9
4	4 × 3	12

    a. Input: number of chairs; Calculation: number of chairs × number of legs on a chair; Output: number of rubber feet for chairs to order
    b. Input: Number of wheels on a tricycle; Calculation: number of tricycles; Output: number of wheels in inventory
    c. Input: number of tricycles; Calculation: number of wheels on a tricycle; Output: number of wheels in inventory
    d. Input: number of booties for dogs; Calculation: number of dogs; Output: number of booties in inventory

36. What is the measure of Angle *B* to the nearest degree displayed on the following protractor?

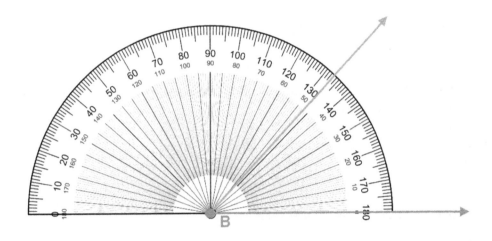

   a. 136°
   b. 133°
   c. 53°
   d. 47°

37. Your friend wants you to help her draw a 110° angle. She has drawn the first ray along the bottom, starting at point *Z*. To complete the angle, you tell her that she needs to draw the second ray starting at point *Z* and passing through which other point?

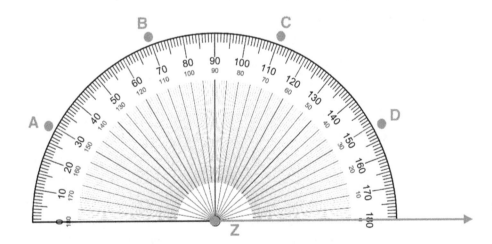

   a. Point *A*
   b. Point *B*
   c. Point *C*
   d. Point *D*

38. A piggy bank contains 12 dollars' worth of nickels. A nickel weighs 5 grams, and the empty piggy bank weighs 1050 grams. What is the total weight of the full piggy bank?
    a. 1,110 grams
    b. 1,200 grams
    c. 2,250 grams
    d. 2,200 grams

39. When rounding 245.2678 to the nearest thousandth, which place value would be used to decide whether to round up or round down?
    a. Ten-thousandth
    b. Thousandth
    c. Hundredth
    d. Thousand

40. What is $\frac{420}{100}$ rounded to the nearest whole number?
    a. 4
    b. 3
    c. 5
    d. 6

41. Round to the nearest tenth 8.067.
    a. 8.07
    b. 8.10
    c. 8.00
    d. 8.11

42. A construction company is building a new housing development with the property of each house measuring 30 feet wide. If the length of the street is zoned off at 345 feet, how many houses can be built on the street?
    a. 11
    b. 115
    c. 11.5
    d. 12

43. The value of 6 x 12 is the same as:
    a. 2 x 4 x 4 x 2
    b. 7 x 4 x 3
    c. 6 x 6 x 3
    d. 3 x 3 x 4 x 2

44. The total perimeter of a rectangle is 36 cm. If the length of each side is 12 cm, what is the width?
    a. 3 cm
    b. 12 cm
    c. 6 cm
    d. 8 cm

45. The following questions are based on this graph of test scores for three students who have classes with four teachers:

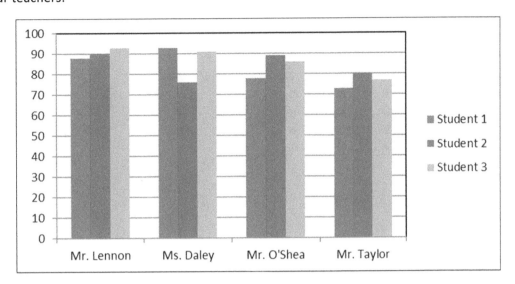

Based on the graph, how many more points did student 3 get on her test in Mr. O'Shea's class than on her test in Mr. Taylor's class?
   a. 5
   b. 8
   c. 10
   d. 14

46. Bernard can make $80 per day. If he needs to make $300 and only works full days, how many days will this take?
   a. 2
   b. 3
   c. 4
   d. 5

The table below shows the number of students in Ms. Jackson' class who play each sport?

Sports Played By Students in Ms. Jackson's Class

Sport	Frequency
Soccer	⦀⦀ II
Swimming	I
Track	III
Baseball	⦀⦀ I
Basketball	⦀⦀ I
Tennis	II

47. Which of the following dot plots correctly represents the data in the table?

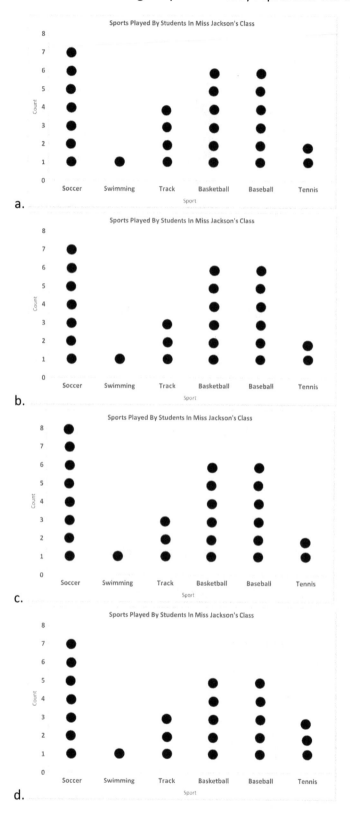

a.

b.

c.

d.

48. Which of the following is an equivalent measurement for 1.3 cm?
   a. 0.13 m
   b. 0.013 m
   c. 0.13 mm
   d. 0.013 mm

49. Karen gets paid a weekly salary and a commission for every sale that she makes. The table below shows the number of sales and her pay for different weeks.

Sales	2	7	4	8
Pay	$380	$580	$460	$620

Which of the following equations represents Karen's weekly pay?
   a. $y = 90x + 200$
   b. $y = 90x - 200$
   c. $y = 40x + 300$
   d. $y = 40x - 300$

50. Jessica buys 10 cans of paint. Red paint costs $1 per can and blue paint costs $2 per can. In total, she spends $16. How many red cans did she buy?
   a. 2
   b. 3
   c. 4
   d. 5

Answer Explanations

1. A: 84,800. The hundred place value is located two digits to the left of the decimal point (the digit 7). The digit to the right of the place value is examined to decide whether to round up or keep the digit. In this case, the digit 8 is 5 or greater so the hundred place is rounded up. When rounding up, any digits to the left of the place value being rounded remain the same and any to its right are replaced with zeros. Therefore, the number is rounded to 84,800.

2. B: The underlined digit is the 6 in 6,000. The bold digit is the 6 in 600. Because 6,000 is equal to 6000×10, we know that the underlined 6 has a value that is 10 times that of the bold 6. Additionally, the base-10 system we use helps us determine that the place value increases by a multiple of ten when you go from the right to the left.

3. D: This solution shows the strip separated into 5 pieces, which is necessary for it to be filled in to show $\frac{3}{5}$. Choice A shows the strip filled to $\frac{1}{2}$, and choice B shows the strip filled to $\frac{2}{4}$, which is also $\frac{1}{2}$. Neither of these selections is correct. While Choice C shows 3 portions filled, the total number of portions is only 4, making the fraction filled $\frac{3}{4}$. This is also an incorrect choice.

4. A: This choice shows that Ming plus his three friends (1 + 3 = 4) is the number of divisions necessary to split the lot of cards evenly (16 ÷ 4 = 4). There would need to be four groups of 4 cards each, or $\frac{4}{16}$, which is $\frac{1}{4}$ of the total cards. The other choices do not correctly divide the cards into even groupings.

5. B: $\frac{1}{2}$ is the same fraction as $\frac{4}{8}$, and would both fill up the same portion of a number line.

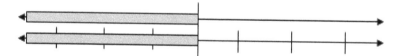

None of the other choices represent equivalent portions to each other, as seen below.

A.

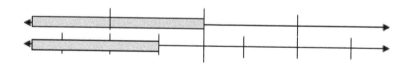

C.

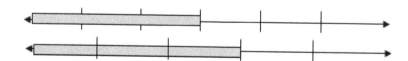

D.

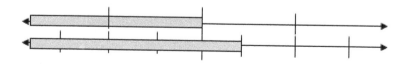

84

6. D: Even though all of the fractions have the same numerator, this is the one that represents the greatest part of the whole. All other choices are smaller portions of the whole, as seen by this graphic representation.

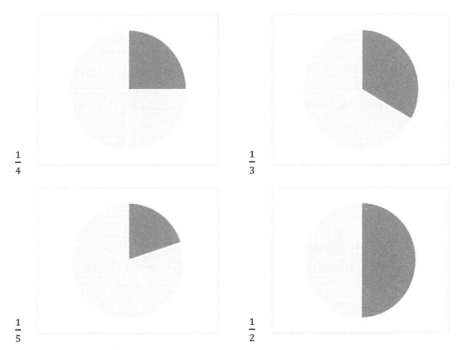

$\frac{1}{4}$

$\frac{1}{3}$

$\frac{1}{5}$

$\frac{1}{2}$

7: A: $\frac{8}{7}$ is the same as $8 \times \frac{1}{7}$, which is represented by the first option: $\frac{1}{7} + \frac{1}{7} + \frac{1}{7} + \frac{1}{7} + \frac{1}{7} + \frac{1}{7} + \frac{1}{7} + \frac{1}{7}$. This can be thought of as cutting a pie into seven slices and then serving 8 slices. Since you need more slices than you have in your pie, you actually need to cut up two pies and take one piece from the second pie. This is because $\frac{8}{7}$ is an improper fraction, which means the numerator (top number) is greater than the denominator (bottom number).

8. A: In order to compare the fractions $\frac{4}{7}$ and $\frac{5}{9}$, a common denominator must be used. The least common denominator is 63, which is found by multiplying the two denominators together (7×9). The conversions are as follows:

$$\frac{4}{7} \times \frac{9}{9} = \frac{36}{63}$$

$$\frac{5}{9} \times \frac{7}{7} = \frac{35}{63}$$

Although they walk nearly the same distance, $\frac{4}{7}$ is slightly more than $\frac{5}{9}$ because $\frac{36}{63} > \frac{35}{63}$. Remember, the sign > means "is greater than." Therefore, Chris walks further than Tina, and Choice *A* correctly shows this expression in mathematical terms.

9. A: The total fraction taken up by green and red shirts will be $\frac{1}{3} + \frac{2}{5} = \frac{5}{15} + \frac{6}{15} = \frac{11}{15}$. The remaining fraction is $1 - \frac{11}{15} = \frac{15}{15} - \frac{11}{15} = \frac{4}{15}$.

10. C: If she has used 1/3 of the paint, she has 2/3 remaining. $2\frac{1}{2}$ gallons are the same as $\frac{5}{2}$ gallons. The calculation is $\frac{2}{3} \times \frac{5}{2} = \frac{5}{3} = 1\frac{2}{3}$ gallons.

11. C: The range of the entire stem-and-leaf plot is found by subtracting the lowest value from the highest value, as follows: $59 - 20 = 39$ cm. All other choices are miscalculations read from the chart.

12. C: To calculate the total greater than 35, the number of measurements above 35 must be totaled; 36, 37, 37, 47, 49, 54, 56, 59 = 8 measurements. Choice *A* is the number of measurements in the 3 categories, Choice *B* is the number in the 4 and 5 categories, and Choice *D* is the number in the 3, 4, and 5 categories.

13. D: $\frac{3}{100}$. Each digit to the left of the decimal point represents a higher multiple of 10 and each digit to the right of the decimal point represents a quotient of a higher multiple of 10 for the divisor. The first digit to the right of the decimal point is equal to the value ÷ 10. The second digit to the right of the decimal point is equal to the value ÷ (10×10), or the value ÷ 100.

14. D: Area = length x width. The answer must be in square inches, so all values must be converted to inches. $\frac{1}{2}$ ft is equal to 6 inches. Therefore, the area of the rectangle is equal to $6 \times \frac{11}{2} = \frac{66}{2} = 33$ square inches.

15. D: The two lines are neither parallel nor perpendicular. Parallel lines will never intersect or meet. Therefore, the lines are not parallel. Perpendicular lines intersect to form a right angle (90°). Although the lines intersect, they do not form a right angle, which is usually indicated with a box at the intersection point. Therefore, the lines are not perpendicular.

16. A: This choice can be determined by comparing the place values, beginning with that which is the farthest left; hundred-thousands, then ten-thousands, then thousands, then hundreds. It is in the hundreds place that Choice A is larger. Choice B is smaller in the ten-thousands place, Choice C is smaller in the tens place, and Choice D is smaller in the ten-thousands place.

17. C: The number line is divided into 10 sections, so each portion represents 0.1. Because the number line begins at 1 and ends at 2, the number in question would be between those two numbers. Since there are only two portions out of ten marked, this represents the number 1.2. All other choices are incorrect due to a misreading of the number line.

18. C: To add fractions, the denominator must be the same. This is the only choice with both denominators of 6. Adding the numerators totals 7, for a fraction of $\frac{7}{6}$. Choice A equals $\frac{7}{3}$, Choice B equals $\frac{7}{5}$, and Choice D equals $\frac{8}{4}$ or 2.

19. A: The number line is divided into ten portions, so each mark represents 0.1. Halfway between the 6th and 7th marks would be 0.65. Choice B shows 0.55, Choice C shows 0.25, and Choice D shows 0.45.

20. A: The light gray portion represents $\frac{4}{5}$ and the dark gray portion represents $\frac{3}{5}$, to total $\frac{7}{5}$. Choice B is not correct because it misrepresents the dark gray portion. Choice C is not correct because it misrepresents the light gray portion. Choice D is not correct because it includes the 1 with the dark gray portion.

21. C: The triangle in Choice *B* doesn't contain a line of symmetry. The figures in Choices *A* and *D* do contain a line of symmetry but it is not the line that is shown here. Choice *C* is the only one with a correct line of symmetry shown, such that the figure is mirrored on each side of the line.

22. C: Perimeter is found by calculating the sum of all sides of the polygon. $9 + 9 + 9 + 8 + 8 + s = 56$, where s is the missing side length. Therefore, 43 plus the missing side length is equal to 56. The missing side length is 13 cm.

23. B: A rectangle is a specific type of parallelogram. It has 4 right angles. A square is a rhombus that has 4 right angles. Therefore, a square is always a rectangle because it has two sets of parallel lines and 4 right angles.

24. C: The situation can be described by the equation $? \times 2$. Filling in for the missing numbers would result in $3 \times 2 = 6$ and $7 \times 2 = 14$. Therefore, the missing numbers are 6 and 14. The other choices are miscalculations or misidentification of the pattern formed by the table.

25. B: This is the only shape that has no parallel sides, and therefore cannot be a parallelogram. Choice *A* has one set of parallel sides and Choices *C* and *D* have two sets of parallel sides.

26. A: The way to calculate the measure of angle 2 is to subtract angle 1 from the measure of a straight line (180^0). $180^0 - 49^0 = 131^0$. Choice *B* subtracts the value of angle 1 from 90^0, Choice *C* subtracts the value of angle 1 from 360^0, and Choice *D* mistakenly labels angle 2 as equal to angle 1.

27. D: In order to solve this problem, the number of feet in a yard must be established. There are 3 feet in every yard. The equation to calculate the minimum number of yards is $79 \div 3 = 26\frac{1}{3}$.

If the material is sold only by whole yards, then Mo would need to round up to the next whole yard in order to cover the extra $\frac{1}{3}$ yard. Therefore, the answer is 27 yards. None of the other choices meets the minimum whole yard requirement.

28. C: To calculate this, the following equation is used: $10 - (5 + 1) = 4$. The number of times the temperature was between 36-38 degrees was 10. Finding the total number of times the temperature was between 28-32 degrees requires totaling the categories of 28-30 degrees and 30-32 degrees, which is $5 + 1 = 6$. This total is then subtracted from the other category in order to find the difference. Choice *A* only subtracts the 28-30 degrees from the 36-38 degrees category. Choice *B* only subtracts the 30-32 degrees category from the 36-38 degrees category. Choice *D* is simply the number from the 36-38 degrees category.

29. D: Area = length x width. Therefore, the area of the rectangle is equal to $3 \times 8 = 24$ square inches.

30. C: First, we need to add up the money Taylor had: 2 $1 bills, 5 quarters, 3 dimes, and 1 nickel:

$$(2 \times \$1.00) + (5 \times \$.25) + (3 \times \$.10) + \$.05 = \$3.60$$

Taylor buys 1 brownie, 2 cookies, 1 cupcake, 2 lemon squares, and 1 container of milk. Using the price list, we can write an equation to represent the total cost of his purchases:

$$\$0.50 + (2 \times \$.20) + \$0.75 + (2 \times \$.60) + \$.35 = \$3.20$$

To calculate this change, we subtract his costs ($3.20) from what he gave the cashier ($3.60) to get $.40.

31. B: 180 miles. The rate, 60 miles per hour, and time, 3 hours, are given for the scenario. To determine the distance traveled, the given values for the rate (r) and time (t) are substituted into the distance formula and evaluated: $d = r \times t \rightarrow d = (60mi/h) \times (3h) \rightarrow d = 180mi$.

32. C: In order to calculate the profit, an equation modeling the total income less the cost of the materials needs to be formulated. $\$60 \times 20 = \$1,200$ total income. $60 \div 3 = 20$ sets of materials. $20 \times \$12 = \$240$ cost of materials. $\$1,200 - \$240 = \$960$ profit. Choice A is not correct, as it is only the cost of materials. Choice B is not correct, as it is a miscalculation. Choice D is not correct, as it is the total income from the sale of the necklaces.

33. D: The slope from this equation is 50, and it is interpreted as the cost per gigabyte used. Since the g-value represents number of gigabytes and the equation is set equal to the cost in dollars, the slope relates these two values. For every gigabyte used on the phone, the bill goes up 50 dollars.

34. D: This problem can be solved by using unit conversions. The initial units are miles per minute. The final units need to be feet per second. Converting miles to feet uses the equivalence statement 1 mile=5,280 feet. Converting minutes to seconds uses the equivalence statement 1 minute=60 seconds. Setting up the ratios to convert the units is shown in the following equation: $\frac{72\ miles}{90\ minutes} * \frac{1\ minute}{60\ seconds} * \frac{5280\ feet}{1\ mile} = 70.4$ feet per second. The initial units cancel out, and the new, desired units are left.

35. C: These labels correctly describe a real-world application of the input-output table shown. The number of tricycles would need to be multiplied by 3 (the number of wheels on a tricycle) in order to find the number of total wheels in a store's inventory. Choice A is not a correct modeling of a real-world situation. A stable chair would have 4 legs, not 3. Choice B is incorrect as it mixes up the number of wheels on a tricycle with the number of tricycles. The number of wheels cannot be the variable (changing) item for this calculation. Choice D does something similar as Choice B, by mixing up the variable and the multiplier; dogs would have a set number of paws, not one that would change.

36. D: Angle B is an acute angle because it is smaller than a right angle, which is 90°. Therefore, we can immediately eliminate Choices A and B. To determine the measure of the angle, look at where the ray that is not along the bottom crosses the arc of the protractor. It falls between the 40 and 50; specifically, it is at 47°. If the ray along the bottom was going towards the left and the other ray stayed where it is now, the angle would be obtuse, and the number of degrees would be read as 133°.

37. B: An angle measuring 110° is an obtuse angle because it is larger than a right angle, which is 90°. Therefore, we can immediately eliminate Choices C and D. Point B is positioned at 110°, so it is the correct choice. A ray through Point A would make a 150° angle, a ray through Point C would make a 70° angle, and a ray through Point D would make a 30° angle.

38. C: A dollar contains 20 nickels. Therefore, if there are 12 dollars' worth of nickels, there are $12 \times 20 = 240$ nickels. Each nickel weighs 5 grams. Therefore, the weight of the nickels is $240 \times 5 = 1,200$ grams. Adding in the weight of the empty piggy bank, the filled bank weighs 2,250 grams.

39. A: The place value to the right of the thousandth place, which would be the ten-thousandth place, is what gets used. The value in the thousandth place is 7. The number in the place value to its right is greater than 4, so the 7 gets bumped up to 8. Everything to its right turns to a zero, to get 245.2680. The zero is dropped because it is part of the decimal.

40. A: Dividing by 100 involves mean shifting the decimal point of the numerator to the left by 2. The result is 4.2 and rounds to 4.

41. B: 8.1

To round 8.067 to the nearest tenths, use the digit in the hundredths.

6 in the hundredths is greater than 5, so round up in the tenths.

8.0̲67

0 becomes a 1.

8.1

42. A: 11. To determine the number of houses that can fit on the street, the length of the street is divided by the width of each house: $345 \div 30 = 11.5$. Although the mathematical calculation of 11.5 is correct, this answer is not reasonable. Half of a house cannot be built, so the company will need to either build 11 or 12 houses. Since the width of 12 houses (360 feet) will extend past the length of the street, only 11 houses can be built.

43. D: By grouping the four numbers in the answer into factors of the two numbers of the question (6 and 12), it can be determined that (3 x 2) x (4 x 3) = 6 x 12. Alternatively, each of the answer choices could be prime factored or multiplied out and compared to the original value. 6×12 has a value of 72 and a prime factorization of $2^3 \times 3^2$. The answer choices respectively have values of 64, 84, 108, and 72 and prime factorizations of 2^6, $2^2 \times 3 \times 7$, $2^2 \times 3^3$, and $2^3 \times 3^2$, so answer D is the correct choice.

44. C: The formula for the perimeter of a rectangle is P=2L+2W, where P is the perimeter, L is the length, and W is the width. The first step is to substitute all of the data into the formula:

$$36 = 2(12) + 2W$$

Simplify by multiplying 2x12:

$$36 = 24 + 2W$$

Simplifying this further by subtracting 24 on each side, which gives:

$$36-24 = 24-24+2W$$

$$12= 2W$$

Divide by 2:

$$6 = W$$

The width is 6 cm. Remember to test this answer by substituting this value into the original formula: 36 = 2(12) + 2(6).

45. A: To calculate the difference between the two scores for Student 3, subtract score from Mr. Taylor's class (77) from the score in Mr. O'Shea's class (85). $85 - 77 = 8$.

46. C: 300/80 =30/8 = 15/4 =3.75. But Bernard is only working full days, so he will need to work 4 days, since 3 days is not sufficient.

47. B: The dot plot in Choice *B* is correct because, like the table, it shows that 7 students play soccer, 1 swims, 3 run track, 6 play basketball, 6 play baseball, and 2 play tennis. Each dot represents one student, just like one hash mark in the table represents one student.

48. B: 100 cm is equal to 1 m. 1.3 divided by 100 is 0.013. Therefore, 1.3 cm is equal to 0.013 mm. Because 1 cm is equal to 10 mm, 1.3 cm is equal to 13 mm.

49. C: $y = 40x + 300$. In this scenario, the variables are the number of sales and Karen's weekly pay. The weekly pay depends on the number of sales. Therefore, weekly pay is the dependent variable (y) and the number of sales is the independent variable (x). Each pair of values from the table can be written as an ordered pair (x, y): (2,380), (7,580), (4,460), (8,620). The ordered pairs can be substituted into the equations to see which creates true statements (both sides equal) for each pair. Even if one ordered pair produces equal values for a given equation, the other three ordered pairs must be checked.

The only equation which is true for all four ordered pairs is $y = 40x + 300$:

$$380 = 40(2) + 300 \rightarrow 380 = 380$$

$$580 = 40(7) + 300 \rightarrow 580 = 580$$

$$460 = 40(4) + 300 \rightarrow 460 = 460$$

$$620 = 40(8) + 300 \rightarrow 620 = 620$$

50. C: Let r be the number of red cans and b be the number of blue cans. One equation is $r + b = 10$. The total price is $16, and the prices for each can means $1r + 2b = 16$. Multiplying the first equation on both sides by -1 results in $-r - b = -10$. Add this equation to the second equation, leaving $b = 6$. So, she bought 6 *blue* cans. From the first equation, this means r = 4; thus, she bought 4 *red* cans.

Photo Credits

Dear Common Core Math Grade 4 Test Taker,

We would like to start by thanking you for purchasing this study guide for your exam. We hope that we exceeded your expectations.

Our goal in creating this study guide was to cover all of the topics that you will see on the test. We also strove to make our practice questions as similar as possible to what you will encounter on test day. With that being said, if you found something that you feel was not up to your standards, please send us an email and let us know.

We have study guides in a wide variety of fields. If you're interested in one, try searching for it on Amazon or send us an email.

Thanks Again and Happy Testing!
Product Development Team
info@studyguideteam.com

FREE Test Taking Tips DVD Offer

To help us better serve you, we have developed a Test Taking Tips DVD that we would like to give you for FREE. **This DVD covers world-class test taking tips that you can use to be even more successful when you are taking your test.**

All that we ask is that you email us your feedback about your study guide. Please let us know what you thought about it – whether that is good, bad or indifferent.

To get your **FREE Test Taking Tips DVD**, email freedvd@studyguideteam.com with "FREE DVD" in the subject line and the following information in the body of the email:

   a. The title of your study guide.

   b. Your product rating on a scale of 1-5, with 5 being the highest rating.

   c. Your feedback about the study guide. What did you think of it?

   d. Your full name and shipping address to send your free DVD.

If you have any questions or concerns, please don't hesitate to contact us at freedvd@studyguideteam.com.

Thanks again!

9 781628 455830